THE ENVIRONMENTALIST'S DILEMMA

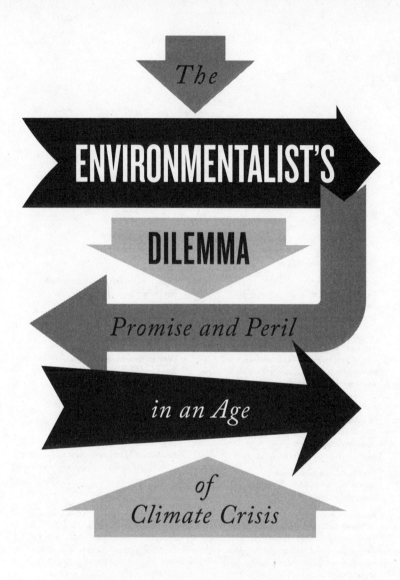

The
ENVIRONMENTALIST'S
DILEMMA

Promise and Peril

in an Age

of
Climate Crisis

ARNO KOPECKY

Published by ECW Press
665 Gerrard Street East
Toronto, Ontario, Canada M4M 1Y2
416-694-3348 / info@ecwpress.com

Editor for the Press: Jennifer Knoch
Cover design: Michel Vrana
Author photo: Christopher Edmonstone

LIBRARY AND ARCHIVES CANADA CATALOGUING IN
PUBLICATION

Title: The environmentalist's dilemma : promise
and peril in an age of climate crisis / Arno Kopecky.

Names: Kopecky, Arno, author.

Identifiers: Canadiana (print) 20210240148 |
Canadiana (ebook) 20210240199

ISBN 978-1-77041-609-3 (softcover)
ISBN 978-1-77305-824-5 (ePub)
ISBN 978-1-77305-825-2 (PDF)
ISBN 978-1-77305-826-9 (Kindle)

Subjects: LCSH: Environmentalism—Moral and
ethical aspects. | LCSH: Environmentalists—Conduct
of life. | LCSH: Sustainable living—Decision making.

Classification: LCC GE195 .K67 2021 | DDC
333.72—dc23

We acknowledge the support of the Canada Council for the Arts. *Nous remercions le Conseil des arts du Canada de son sout-
ien.* This book is funded in part by the Government of Canada. *Ce livre est financé en partie par le gouvernement du Canada.*
We acknowledge the support of the Ontario Arts Council (OAC), an agency of the Government of Ontario, which last
year funded 1,965 individual artists and 1,152 organizations in 197 communities across Ontario for a total of $51.9 million.
We also acknowledge the support of the Government of Ontario through the Ontario Book Publishing Tax Credit, and
through Ontario Creates.

ONTARIO ARTS COUNCIL
CONSEIL DES ARTS DE L'ONTARIO
an Ontario government agency
un organisme du gouvernement de l'Ontario

Canada Council   Conseil des arts
for the Arts   du Canada

Canada

PRINTED AND BOUND IN CANADA   PRINTING: MARQUIS   5   4   3   2   1

MIX
Paper from
responsible sources
FSC   FSC® C103567
www.fsc.org

For Kiran

CONTENTS

THE NEWEST NORMAL

TREES, WE KNOW, ARE GOOD. They support ecosystems and economies, scrub the sky clean with their leaves, and hold the earth together with their roots. This is especially valuable on steep mountainsides like those that rise above the city where I live. Vancouver used to be a logging town; the local forests were liquidated long ago, but the jagged range that forms our northern border flaunts an embarrassment of second-growth riches. I gaze at those green slopes every day from the room where I write. I'm staring at them now. I'm thinking how, despite trees' many fine qualities, the easiest thing to picture them doing is the one bad thing they can: catch fire.

Trees have been doing a lot of that lately. From Australia to Siberia, the Pacific Northwest to the Amazon, this century has been one of ever-bigger forest fires; the flames haven't yet reached Vancouver, but the smoke sure has. It blows in most summers now, usually in August or September when wildfire season hits its stride, hiding our mountains behind a grey veil and reminding us how quickly a symptom can become a cause. The forests of British Columbia have gotten so flammable over the past twenty years that they've started to emit more carbon dioxide than they absorb. Hotter, drier summers are, of course, a factor, but a more poetic influence on our Air Quality Index is the way a hundred years of fire suppression has yielded the largest fires in recorded history. It used to be that Indigenous communities throughout North America and much of the world lit small controlled burns each spring and fall to remove the undergrowth that fuels these conflagrations, but we settlers suppressed those nations and their practices, too. Now here we are, on the cusp of a new, fifth season that lands sometime between summer and autumn, and not just in Vancouver.

Unpredictable yet increasingly consistent, smoke season dominates our conversations whenever it arrives. It's novel, and ominous, and fleeting, the perfect conversational kindling. *Imagine*, we say, *we're breathing in what used to be a forest, animals included*. Or simply, *What a sunset*.

More people are dramatically impacted by forest fires each year, through power outages, evacuation orders, incinerated homes, or even death. But it's also true that for most of us life continues pretty much as usual during smoke season, just with more sore throats and fewer walks. We close the windows for a week or three, until the wind shifts and the sun grows bright again. Then we return to our distractions, busy citizens of wealthy countries with no time to waste on things we can't control. Gradually, we forget about the strange predicament so briefly illuminated by that all-obscuring smoke: Things have never been so good for humanity, nor so dire for the planet.

↞∿↠

Consider the ledger.

On one side is everything from democracy and the global spread of literacy to modern dentistry, plumbing, and the light bulb. My wife happens to deliver babies for a living, a realm that offers a profound illustration of historical improvement. Pregnancy and childbirth killed roughly one of every hundred mothers prior to the Industrial Revolution; today it's one in ten thousand throughout the industrialized world. In Canada, a century ago, one in five children died before their fifth birthday; that's down to one in two hundred today. For all those countries on Earth where mortality remains too high, change is coming fast. Between 2000 and 2017, global maternal mortality dropped by 38 percent, while under-five child mortality was cut in half.

This kind of progress isn't limited to reproductive health. You can track similar trends on practically any quantifiable aspect of human well-being. Gender equality, food security, and public education are spreading round the planet, as are access to medicine and the whole spectrum of a recent invention called human rights.

On the other side of the ledger is all the life that isn't ours. For as long as our species has been improving its own lot, we've been executing a simultaneous campaign of annihilation known as the sixth great extinction. A comprehensive study published in the *Proceedings of the National Academy of Sciences* in 2018 found that human civilization has so far wiped out 83 percent of the world's mammals, half the plants on Earth, and 15 percent of the oceans' fish. In 2019, a United Nations report warned that up to one million species face extinction this century, which is to say in the probable lifetime of most kids born today. This is both a moral abomination and an existential threat, because the same forces wiping these species out will eventually come for us.

The moral dimension of our assault on the biosphere is a hard thing to contemplate. I mean that literally — it is difficult to notice. Non-human suffering occupies a spectrum of moral light our eyes struggle to register. After all, how many times throughout history have humans justified the murder or enslavement of others by portraying them as animals? Empathy does come more easily as the animals grow bigger, more intelligent and expressive, or just cuter. But even for the most charismatic megafauna, our sympathies are fickle. I myself don't spend much time thinking about how lonely the last orangutans of Borneo must be, and I probably think about them more than most.

Still, it does frequently happen that some wild species' plight grabs hold of our imaginations. In the summer of 2018, one such story unfolded in the waters to my west. It began when a killer whale known as Tahlequah gave birth to a calf that died half an hour later; overcome by grief, Tahlequah refused to let her baby go and instead carried it with her, raising the body above the surface as though to help it breathe, over and over again, for seventeen days.

Seventeen days. For people all over the world, Tahlequah's display caused a double jolt: the familiar one of a mother's grief and, more profoundly, the unexpected sight of our reflection in a whale. No one was astonished to learn that orcas are struggling to survive in industrialized waters (it would be surprising if they weren't), but the prolonged bereavement of an *animal* so clearly deranged by grief did something that anyone who's ever tried to write a story or start a movement understands the value of: It turned knowledge into feeling.

Everyone knows that our oceans and the creatures that swim in them are in trouble. Everyone knows whose fault that is. Humans didn't kill Tahlequah's child directly, but we are very much the reason why her community — the Southern Resident orcas, who ply the coast between Seattle and Vancouver — is on the brink of extirpation, reduced to fewer than eighty individuals as of this writing. Tormented by whale-watchers and pollution and crashing salmon populations and a degree of acoustic agony no human can fully comprehend, the Southern Residents' suffering is both a tragedy and a cautionary tale.

And this is where the moral disaster becomes an existential crisis. Even if your cold, anemic heart is unmoved by the Tahlequahs of the world, there are perfectly selfish reasons to protect them and their habitat. Forget about killer whales. An ocean with more plastic than fish won't be an endless source of protein. Slaughtering the world's pollinating insects isn't a great agricultural strategy. Throw in the global depletion of potable freshwater; hyper-volatile weather bringing ever more droughts, fires, floods, and hurricanes; plus, oh I don't know, rising sea levels set to displace one or two billion coastal inhabitants before the end of the century, and it all becomes — like nothing else but nuclear war — too much to contemplate.

Has it ever been easier not to? The grocery stores in which I've foraged all my life suggest ever more abundance and diversity from one year to the next. That message, and countless others like it, hits me on a far more visceral level than any data-driven communiqué from the United Nations Framework Convention on Climate Change. Today's music is so excellent, the television so sophisticated, the internet so bewitching that it's harder than ever to feel the moral, mortal peril that we're in.

It's like we've turned Noah's ark into a humans-only party yacht and sailed it to the edge of Niagara Falls. There are a million distractions aboard, but only three options as far as the waterfall goes. You can struggle against all odds to turn the ship around, stare numbly into the abyss, or turn your back and dance.

My personal adaptation is to ricochet erratically between all three. But on those days when I'm going with the first one, standing on the dance floor shouting, *Guys, guys*, I take a certain solace from the fact that, lately, more of us are waving our arms.

〜〜〉

I know. There's a caveat in the room.

If I were a coltan miner in the Democratic Republic of the Congo, or a Syrian refugee, or a survivor of Canada's residential school system, I wouldn't be going on about how fabulous life is.

There's a terrible danger in praising human progress, one that white men like myself are embarrassingly prone to: We mistake our good luck and the tireless work of others for personal merit, and we promptly forget about the multitudes who remain mired in desperate circumstances. There are still some 700 million undernourished people in the world, plus eighty million refugees and internally displaced peoples. Misery and injustice aren't confined to the developing world, either. Over 50 percent of the kids in Canada's foster care system are Indigenous, while the proportion of incarcerated Americans who are Black is three times that of the general population. There is no end to statistics like these or the stories embedded within them. How can we worry about polar bears while so much human suffering remains?

A subconundrum then: Not only are things too good for us to worry about the environment, they're also too bad.

But there's a commensurate danger in dismissing the hard-fought gains of progress: They vanish when taken for granted. Every good thing in this world requires maintenance, from love to public infrastructure. We're seeing today how easily things like racism and the measles can creep back into societies from which they'd supposedly been eradicated. The U.S. has even allowed maternal mortality to start inching back up, from seven deaths per hundred thousand in 1990 to over seventeen today (a figure that rises to forty-two for Black women). That's awful; it's also still a hell of a lot better than a hundred years ago, when the number was over six hundred.

For much of human history, slavery was endemic to every continent. Women were denied the vote, if there was one. We beat our children. We casually deployed violent slurs against those of different race or religion or sexual orientation and marginalized them to death. Slowly, fitfully, with all kinds of backsliding and failures of principle and mountains yet to climb, this is changing. If you doubt this, try reading a newspaper from 1925, or 1948, or 1976, and see how national leaders and everyday citizens from across the political spectrum described women, or Black and Indigenous people, or people from other countries and religions.

Are there innumerable searing human crises in the world today? Yes. There always have been, along with the moral obligation to alleviate them. What there hasn't always been is an international standard of human rights, motorized transport, or elected government. There's never been eight billion humans trying simultaneously to preserve their identities and merge into a unified global society. And there has never in the lifetime of our species been the prospect of a planetary ecological collapse.

"This is an extraordinary time full of vital, transformative movements that could not be foreseen," Rebecca Solnit wrote in 2015. "It's also a nightmarish time. Full engagement requires the ability to perceive both."

Truer every day. So I find myself arguing, in all the wrong places, to all the wrong people, that things have never been better! But we should be worried, very worried!

↞∿↠

It's not like I'm the first to suffer this condition. In 2010, *Bioscience* magazine even gave it a label: the Environmentalist's Paradox. In an article called "Untangling the Environmentalist's Paradox: Why Is Human Well-Being Increasing while Ecosystem Services Degrade?" a multidisciplinary team of researchers examined the problem through the lens of four hypotheses:

1) Maybe humans aren't actually better off, but only think we are.
2) Advances in food production outweigh all other considerations.
3) Modern technology has reduced our reliance on ecosystem services.

4) The worst effects of environmental degradation are
   yet to come.

The researchers wound up rejecting the first hypothesis outright and accepting the other three with caveats. It's no mirage:
We really are doing better than our ancestors. A big part of that
improvement comes down to having more and better food, which
takes much of the sting from downsides like disappearing forests.
In addition to agricultural technology, other technical innovations
have reduced our species' dependence on local ecosystems, so that
(for example) when the river you drink from grows polluted, it's
possible to install a water treatment plant. But in spite of these
protections, the time lag between cause and effect in planetary
systems — such as the delay between burning fossil fuels and glaciers melting — means that the full consequences of our actions
are only starting to make themselves known.

In other words, the Environmentalist's Paradox is no paradox
at all but a simple matter of living on borrowed time. The worst
is yet to come.

Ten years after that paper appeared, this conclusion seems
more obvious. It's remarkable, really, how much the global mood
has darkened in the past decade. In 2010, we'd just overcome a
global financial crisis. The Bush administration, whose tenure felt
like a fossil fuel conglomerate stole the keys to the world's biggest
nuclear arsenal, was finally gone, replaced by an Obama administration that at least talked a good game on things like climate
change and empire and which for all its flaws was a reliable proponent of multilateralism, honesty, and expertise. In 2010, the
world's problems felt real but surmountable, and the burden of
proof in any argument between human prosperity and ecological
collapse still fell to the latter. But now, after a decade of mounting calamities, the tables have turned. The presidency of Donald
Trump was as hard on advocates of human progress as it was on

environmentalists. It's too early to know whether those four interminable years represented a temporary backlash or an irreversible tipping point, but it's fair to say that by the time 2020 limped to a close, the case for rosy predictions of any sort was forced at best.

But shifting from blithe optimism to knee-jerk pessimism doesn't help anyone escape the Environmentalist's Paradox. If you focus solely on forest fires or COVID-19's death toll, you forget (and endanger) the many positive developments of this century's teenage years. The #MeToo movement and Black Lives Matter and Canada's reckoning with Truth and Reconciliation; the entry of a billion or so people into the middle class; the unabated rise to power of women throughout the world, including the United States' first female, Black vice-president: these and many other good things happened, too.

"Remembering," the renowned British environmental writer George Monbiot has said, "is a radical act." He was talking about humanity's inability to perceive incremental change, which Monbiot regards as one of our most dangerous blind spots and which scientists refer to as the phenomenon of "shifting baselines." That term is usually invoked in the sense Monbiot meant, to describe how each generation grows accustomed to a diminished ecosystem and fails to register that anything might be missing. In this way, we never realize that we're catching fewer and smaller fish than our parents, or that there's nowhere near as many bugs as before; our conception of biological abundance is constantly being downgraded without anyone noticing. But baselines shift upward, too: I can fly to Paris, treat an infection with penicillin, or pluck an ice cube from the freezer without any of these things seeming remarkable. The baselines of material progress no longer take a whole generation to shift up, either. It already requires conscious effort to be amazed at what my phone can do or how normal gay marriage now seems.

Remembering is radical because it cuts against the grain of our capacity to adapt. That's one of humanity's most celebrated traits. Adapting is how our species spread to every biome on the planet, how we learned not merely to survive in harsh environments, like the Arctic, but to love them so fiercely that the people who came to call them home would rather die than leave. Adaptation is one of our most celebrated traits, and rightly so. It's hard to survive if you can't adapt, if you veer constantly between wonder and horror and exist in a perpetual state of astonishment. We have a word for people like that: children.

But adapting necessarily includes a measure of forgetting, and that, too, has become a problem. Because any hope of escaping the Environmentalist's Paradox rests, it seems to me, on all of us learning to be radical. We must simultaneously remember how far we've come and how much we've lost.

The alternative is to remain numb, to stay used to it all. One day, the Great Barrier Reef is reduced to white skeleton; the next, a cure for Alzheimer's is announced; the news that three billion birds have disappeared from the skies of North America is followed by the news that America's cabinet appointments are the most diverse in history. These things flare up, delight or horrify everyone for a news cycle, then become that most modern phenomenon: the new normal.

❦

New normals have been adding up for so long now that they've gone meta. Our constantly shifting baselines have themselves become a grand new normal.

At least this one has a name: the Environmentalist's Paradox. Which is really no paradox at all. It's just a problem. A big one, for sure, but like any problem it can only be solved if you look squarely at it.

This book is about that gaze. The essay you're almost finished reading, a version of which appeared in the *Globe and Mail* in the summer of 2018, sprang from my belief that our newest normal has become a fundamental source of dissonance in our daily lives, a constant background jangle that corrodes public discourse and poisons our politics precisely because so few of us give it any thought. Instead, we focus on one or the other of its two conflicting halves — on *our* truth — then wonder how we came to be so polarized. From Parliament Hill to the family reunion, how many of our arguments boil down to some version of this single disagreement? A simple but bitter dispute, pitting those concerned with preserving (or reviving or expanding) human prosperity against those who fear for the rest of the biosphere.

In 2018, my aim was to clarify the nature of the dispute. It seemed to me then, and still does today, that a great many environmental advocates — myself included — are as oblivious to our own blind spots as oil barons and mining magnates are to theirs. Since then, I've spent a lot of time wondering, *What if everyone agreed?* What if the industrialists said, *Hey, you're right, the world is finite after all*, and the environmentalists said, *Yes, but people do have to eat?* Would the solutions suddenly be easy?

Ha. The moment you acknowledge that human well-being is on a collision course with environmental collapse, you enter a realm of double-edged questions and difficult decisions. Our governments haven't yet managed to rise to this occasion — should we topple them or try to change them from the inside? Since we're talking about democracies, maybe it's just a matter of injecting the proper urgency into the electorate — but how, exactly, when said electorate is already overwhelmed with bad news? What if we compare climate change to Nazi Germany? How do you tackle mass delusion? What about more personal matters: Should we try to empathize with people we regard as enemies? Is compromise essential or fatal? Should we still be having kids? If so, should we still take them to Disneyland? Does anything we do as individuals even matter?

In the essays that follow, I've followed these questions wherever they wanted to go: to the dawn of the environmental movement, and to Alberta's tortured love affair with oil; to postwar Germany's reckoning with the past, and to post-Trump America's relationship with the future. Wherever, whenever they led, these paths kept returning to an age-old question: How can we engage with the story of our times?

If you keep reading, you'll notice a pandemic swept across the world while I wrote. You might fairly ask, Don't the mass death, loneliness, and economic ruin caused by this novel coronavirus undermine my belief that things have never been so good for humanity? Quite possibly, yes. But this is a book that finds potential in seeming contradictions. I don't want to downplay the tremendous suffering so many have endured, but I do have to ask: When was the last time we developed a vaccine in the same year a pandemic emerged? More importantly, when was the last time global civilization had an opportunity to alter course on such a fundamental level?

Whether we seize that opportunity, and who does the seizing, remains to be seen. But it feels essential to note that never have so many people, from so many walks of life, agreed the status quo isn't working. Whatever normal — the *old* normal — was doing right, it also brought us COVID-19 and the Trump administration. As we emerge from those calamities, blinking amid the wreckage and bracing for whatever might come next, a new current of big thinking seems to have entered the halls of power. The leaders of just about every country on Earth have learned they could rewrite the rules of society. However costly the lesson, it suggests that we could design the next new normal.

It wouldn't be the first time. Previous examples include the five-day work week, public schooling, and the absence of small-pox — but also factory farms, the Indian Act, and offshore tax shelters. The intentional creation of new normals has always been freighted with promise and peril, each rising in proportion to the challenge. For things to get better instead of worse, we'll have to come together in ways that are hard to imagine in the current cultural moment, when the outbreak of mass violence feels as likely as a new era of collaboration. But it seems to me that any hope of overcoming our furious divides lies in embracing our contradictions and running to, not from, the dilemmas they reveal.

Speaking of hope, here's one small crumb to consider before you turn the page. That killer whale, Tahlequah? Two years after she lost her first baby, she gave birth to another. This one, a healthy male, survived. He's swimming somewhere to the west of me as I write. So far as I know, no one's yet determined his life expectancy.

MICKEY MOUSE IS ALL RIGHT

WE DIDN'T HAVE TO GO TO DISNEYLAND. We had no business in Los Angeles. But we wanted to go, and we could, so we did, and maybe under all the smart analysis that's the simple truth of our self-imposed predicament.

How lovely it would be to leave it at that: go on enjoying whatever we can, because we can, while we can. To wait until we can't before we start to worry. Or better yet — to just hold back, slow down, and even stop.

Anyway, we went to Disneyland.

⟨∿⟩

It was the middle of October and our daughter had just turned four, a number she was intensely proud of without having the slightest idea what it meant beyond presents and cake and a vague sense of development. She knew she could do things now that she hadn't been able to before, like touch the bottom of the pool, and that soon she'd be able to do even more, like swim unaided. At this stage of her trajectory, it was inconceivable to her that progress could lead anywhere but up, to anything but better.

That the year was 2019, that in this autumn of Ada's fourth birthday her country would hold its forty-third federal election, her father would turn forty-three, and the carbon dioxide censors on Mauna Loa would surpass 410 parts per million — these numbers were a foreign language. They marked a different kind of progress, saturated with varying degrees of foreboding, and I was glad they meant nothing to her. Glad to let her eat cake and open presents and meet Mickey Mouse, whose own father — the Walt Disney Company — had, after ninety-six years of childlike progress, posted fourth-quarter earnings of $19 billion.

←∿→

Ada didn't yet know about Mickey Mouse. He wasn't a part of our lives back home, though he'd been very present in my wife's and my childhoods. Tradition was at play here — something had delighted us when we were tiny tots and left us with lifelong happy memories; now we had an opportunity to pass it down the line. To share and connect. I tried telling Ada about Mickey outside the car rental agency while my wife dealt with the paperwork inside, but we'd just spent four hours sitting in a plane and Ada was not in a mood to listen. She wanted to dance and spin circles until we fell over. So we danced and spun and fell, then lay on our backs on the concrete just outside the world's fourth busiest airport, directly under the landing path of jet plane after jet plane, roaring by so close you could reach out and tickle their tums.

The sun was going down. Every two minutes another leviathan lumbered over us, impossibly slow. Their blinking lights formed a string of twinkling stars that stretched into the darkening eastern sky. Roar and tickle, scream and giggle, repeat. Ada noticed how the air beneath the jets "looks like water." Eventually my wife emerged with the keys to a Ford F-150 that we hadn't reserved but was all the company had left; we pulled out of the parking lot feeling guilty to be in such a guzzler, but not quite guilty enough to go through whatever we'd have to go through to be driving something else. The sun now gone, we entered the red river of freeway that perforates greater Los Angeles, six lanes of brake lights accordioning from high to low velocity and back again, me feeling like a country rube as I clutched the wheel and tried to concentrate on Siri's directions above Ada's angry protests at being forced to sit down once more, for a whole hour more, as we on- and off-ramped our way to her aunt's apartment.

It is hard, in those circumstances, to properly appreciate the parade of miracles that enable a family to wake up in one city and go to bed in another, two thousand kilometres away, without missing a meal or needing to look at a map.

What's easier, at least for those of a certain bent of mind, is to perceive the carbon emissions that such a trip entails. You can hear, and see, and smell, and finally start to *feel* the fumes of all those internal combustion engines and airplane jets, on display here like almost nowhere else on Earth, endlessly burning and turning the air into water. As many planes as stars, rivers and rivers of cars, and you wonder, *How long can this go on?* How much oil is in the ground, how much carbon can the sky absorb, how fast will we go before we finally slow down?

↞∿↠

Questions like these are not allowed in Disneyland. You leave them at the entrance, along with any alcohol, narcotics, or firearms you might have accidentally left in your bag. This city within a city, literally a walled fortress manned by snipers and populated by armed undercover officers (Disneyland is considered a prime terrorist target), is a place for childlike wonder.

"What I want Disneyland to be most of all is a happy place," Walt Disney once said. "A place where adults and children can experience together some of the wonders of life, of adventure, and feel better because of it."

I'm afraid the vacant grins and the ruthless sun and the kid-sized metal seats had the opposite effect on me. But as anyone who has taken their children here knows, the place casts a happy spell on children — even those, like Ada, who arrive innocent of the Disney-verse, have never seen or heard of Mickey and Minnie and Goofy, Jasmine and Ariel and Elsa. A child can meet them all here for the first time and be instantly enchanted. If all goes as Disney plans, that enchantment will follow you home and stay with you for life.

The first ride we took Ada on was It's a Small World. It had the shortest lineup and, housed inside a large pavilion, was blessedly hidden from the sun. We boarded the sixteen-person boat and commenced our gentle float through the winding interior, taking in the kaleidoscopic succession of hundred-foot dioramas that depict regions of the world, each one populated by animatronic dolls singing the title song, over and over again, in their respective languages. There were children dressed in turbans and head scarves, children dressed as Canadian Mounties and British royal guards, Thai dancers and hula-skirted dancers and cancan dancers, Dutch children in tulip fields, Indian snake-charming children, children in kimonos, children on camels, children in sombreros. There were bush children in loin cloths and children in lederhosen.

They sang to us in English and Spanish and Mandarin and Hindi. It's a small world, they reminded us. Who could disagree? Not me, not my wife, certainly not our four-year-old daughter, who sat riveted from start to finish, rubbernecking for fifteen minutes straight as we glided beneath igloos and toucans and giraffes and hot air balloons. Her delight, inevitably, was infectious. We grown-ups are so accustomed to newness and innovation, to anything that is more than a few years old seeming hopelessly dated, that it was almost unbelievable to behold an attraction built in 1964 maintain its spell-casting power, three generations later. The dolls bob and wave woodenly like characters in a pull-tab book, and the audio is tinny and shrill. But to Ada and millions of other children, it is pure timeless magic.

As we exited into the sunlight, my world bifurcated once more. In one dimension, I was grateful for my daughter's joy. That joy was unimpeachable. I wanted her to lose herself in the ecstasy of the moment, the day, that fleeting period of her life when ignorance really is bliss and not yet an exploitable blind spot.

Because that's what it is in the other dimension, the one I can never quite shake. Outside this benign fantasyland, a much darker fantasyland has displaced reality for tens of millions of Americans. In that reality, Bill Gates and George Soros are the villains, vaccines are evil potions, Cruella de Vil has turned into Hillary Clinton, and Prince Charming is Donald Trump, come to wake America from her slumber with a pussy grab. These Americans would like nothing better than to wall off their fantasy and lock out the Aladdins; they have their own guns, though, so no need for snipers.

Perhaps more frightening than their world of alternative fact is the system of values it was built to protect. In the world's most affluent nation, an untold number of citizens have come to see power as justice, empathy as weakness, confidence as truth. In this world, complexity is cause for suspicion, diplomacy is a waste of time, and climate change is just another lefty hoax. Americans are not alone in any of this, aren't even the most afflicted, but they are the most powerful, and so their dark fantasies inflict more damage on our small world than anyone else's.

Many of the same forces were at play in Walt Disney's time, of course. He and his colleagues were well aware of them, and even motivated by them. It's a Small World debuted at the 1964 New York World's Fair, two years after the Cuban Missile Crisis.

It was, and remains, at least partly a response to the disastrous missteps of grown-ups, "a tribute to the children of the world," as the pamphlets said at the time. The legendary Sherman brothers, who wrote many famous songs for Walt Disney, described this one as "a prayer for peace."

The world outside these walls is a lot more peaceful than it was in 1964, but the potential for fresh violence is increasing once again, not least in Walt Disney's own country. Do children need to know this? Is it wrong to shelter them awhile, let them bask in the illusion — some might say aspiration — of global harmony? It's not as though forcing children to sit through a ride portraying all the world's atrocities would help anyone, though it did occur to me that It's a Small World would make a fine prison for a select handful of world leaders.

It also occurred to me, as I roasted in that sun, that all cultures tell themselves fantasy versions of who they are; that America has done this more effectively than most; that in the years since It's a Small World began singing its prayer of peace to untold millions (Disneyland alone sees nearly twenty million visitors a year), Americans dropped over seven million tons of explosives on Vietnam, supported murderous dictatorships in every country south of Mexico, and turned the Middle East into a perpetual war zone so as to keep the cost of gasoline below a dollar a litre, to say nothing of what it had done and continues to do to its own Black and Indigenous citizens. Almost as bad as those actions is the number of Americans who remain oblivious or indifferent to them, who see their country as a shining meritocracy where words like "genocide" and "slavery" and "extinction" have no place.

Is there a connection between the seemingly harmless fantasies we tell our children and the dark lethal fantasies we sell ourselves as adults? I don't know. Many of the people who contributed to those atrocities might have wandered through Disneyland when they were kids, giggling and eating too much ice cream just like Ada did that day, and I could construct an argument about the malign influence of stereotypes peddled by one Disney franchise after another — the hot royals, the savage hordes (go back far enough and you can watch Mickey prancing around in blackface). But honestly, it's a bit of a stretch from *Aladdin* to the Iraq War. And if we're going to blame Disney for seeding pernicious stereotypes and impossible expectations, shouldn't we also give credit for more positive tropes like kindness and fair play, forgiveness and redemption?

What I know is that fantasies are important, and not just for kids. Adults need dreams, too. Walt Disney understood this very well. "I can't believe there are any heights that can't be scaled by a man who knows the secret of making dreams come true," he said. That secret could be "summarized in four Cs. They are Curiosity, Courage, Confidence and Constancy. And the greatest of all is Confidence. When you believe in a thing, believe in it all the way." Like so much of what Disney said and did, this pleasant-sounding fantasy can all too easily be turned to villainous ends. Let enough grown-ups fall under the spell of Disney's favourite C and you wind up with a confidence man like Donald Trump in control of the nuclear codes.

Reader, let me tell you: There were a lot of adults without children at Disneyland that day.

↞∿↠

It was around this time that my wife informed me I had a severe case of resting bitch face. From then on, I did my best to lift my lips as we took Ada from ride to ride, with ice cream in between. We piled into the spinning tea cups, we flew on the space ships, we drove the bumper cars, and nary a word from me about the cruel consequences of the internal combustion engine. A marching band went by, and I said nothing about the martial implications of their drumbeat; a gaggle of costumed Disney characters appeared, and I felt sorry for the poor saps who either had to dress like sexy princesses or don furry costumes in the ninety-degree heat, but I kept that to myself as well.

We decided it was important for Ada to meet Mickey Mouse. She herself expressed no such desire. But she was happy to accompany us into his home, and so we followed the meandering hallway to the antechamber, where a few dozen kids and their parents were waiting ahead of us. A small movie screen entertained the queue with old Mickey Mouse cartoons. Once again, Ada was mesmerized. The first cartoon showed Donald Duck, Mickey, and Goofy fleeing a malevolent butcher who'd lured them into his shop; he chased them into the street with his butcher knife held high above his head, but they ducked inside a parked truck and gave him the slip. Lo and behold, it was an ice cream truck! With nobody inside it! In the orgy of consumption that ensued, Goofy knocked the parking brake loose without anyone noticing, and the truck began to roll. They must have been in San Francisco, because those were some steep hills. Next thing you knew, Goofy and Mickey and Donald and all the ice cream pails were flying about like astronauts in an anti-gravity chamber, while the truck careened around the city. It was a pretty accurate depiction of American foreign policy. Maybe they weren't sheltering these kids after all.

And then the final door opened. It was our turn to meet the crown prince of this capitalist paradise of emotional manipulation. The Mouse himself. We stepped over the threshold. Ada — who a day ago had never heard of Mickey, was by now an enormous fan, felt as though she'd been waiting all her life to meet him — was as thrilled as I'd ever seen her to meet anyone. She hugged him, held his big white hands in hers, and peered into those big black soulless eyes. For this moment of ultimate, innocent ecstasy, bequeathed by a corporate logo upon my unsuspecting daughter, I forgave the Walt Disney Company for its lifelong-consumer-creating ethos, forgot its obscene profits and ruthless market domination, accepted that Walt Disney was not the first visionary to create an institution whose inheritors fell in thrall to market forces, and let myself bask in a moment of parental joy.

↞∿↠

Two days later, the Santa Ana winds picked up. It was our last day in L.A., and we drove to Venice Beach in our oversized truck. As we rolled down the freeway, we saw smoke rising from the northern hills. Ten minutes later, an alert popped up on my phone: the Pacific Palisades were being evacuated. That neighbourhood was ten kilometres away from Venice Beach, where we parked as though nothing was wrong, because in Venice Beach nothing was wrong. We helped Ada out of the car, strolled around, bought a couple T-shirts, and made pleasantly vapid small talk with the good-looking hipsters minding the store. The subject of fire did not come up.

But, of course, California kept burning, the climate kept on changing, and finally the question couldn't be evaded: Was my daughter's smile worth the jet fuel we burned to earn it?

Our trip's contribution to the planet's atmospheric carbon dioxide was minuscule. According to Shame Plane, a Swedish website that translates air passenger miles into Arctic ice loss, our three round-trip tickets between Vancouver and Los Angeles melted a combined total of 5.4 square metres of the Arctic. (Sweden has quietly become the epicentre of the flight shame movement — flygskam in Swedish — thanks to one of Sweden's most famous citizens, Greta Thunberg, who refuses to travel by air.) Five and a half square metres of ice isn't quite nothing, but it's close enough that I can view my dilemma more as a thought experiment than a moral abyss.

Most children take to sugar with unadulterated joy. No matter how often they're told ice cream is bad for them, offering them a bowl of it yields no mixed emotions. Over time, the relationship gets more complicated. Consequences accumulate. By the time we're adults we might have replaced sugar with any number of ready substitutes, but most of us have experienced the emotional twinge that accompanies pleasurable acts of knowing self-destruction. Take it far enough and even the slightest taste — a bite of chocolate, a sip of wine — can acquire morbid connotations.

My relationship with climate change has followed a similar trajectory. For much of my life, I travelled widely for both pleasure and work without giving much thought to my personal emissions, which, in the larger scheme of things, were infinitesimal. But with the accumulation of time, and age, and climate catastrophes both near and far, the connection between my behaviour and climate change has grown easier to feel, harder to rationalize. Each puff of greenhouse gas emitted on my behalf contributes to everything from wildfires in the Amazon to the terminal decline of Vancouver's Fraser River sockeye salmon run. Unlike eating too much ice cream, for which the consequences are mostly personal, the damages inflicted by my climate sins land almost entirely upon others — notably my daughter, the very person I aimed to please by flying us to California.

I'm talking, of course, about climate guilt. I'm wondering if it's even possible for a climate-conscious individual to participate in an industrialized society like Canada's without incurring some. That's true even if you don't fly to Disneyland with your family, a guilty pleasure at best that no honest appraisal can justify in climate terms. Unfortunately, embarrassingly, to live above the poverty line involves climate-unfriendly actions almost by definition. We drive cars with internal combustion engines, or use electricity derived from fossil fuel combustion, or shop imperfectly, for reasons that range from self-indulgence to necessity. Do you use the internet? If so, you're contributing to an industry that accounts for over three percent of global emissions, comparable to the global airline industry.

The closer you look, the more it becomes a question of where to draw the line. Say we never go back to Disneyland, but my daughter has an aunt in Los Angeles — is it okay to visit her? What if she was sick and needed help? What about Ada's grandparents in Edmonton and Montreal — are we allowed to visit them, or vice versa? If so, how often? What about the very act of having a child? By some accounts, that alone is the worst thing you can do for the climate, with or without Disneyland. Are we to understand parenthood itself as the ultimate climate sin?

There's no point in looking for definitive answers to these kinds of questions. But there is a point in asking. That point is to make ourselves aware of the world we're moving through, to feel the impact of our actions and inactions. The point is to start *noticing*. Some of us will shrug and carry on, allow ignorance to harden into indifference. But I think the more common human reaction is to pause and reflect on the options available to us for changing course. Perfection isn't one of them; perfection is a fantasy, too. But improvement? Doing a little better, every year? In a world of inescapable complicity, that may be the closest we grown-ups can come to the innocence we so adore.

Imminent, at the speed of people, is too late.
— Richard Powers, *The Overstory*

THE VELOCITY OF PERCEPTION

I HAVE DEVELOPED AN OBSESSION with slow-moving phenomena. Things that advance at a crawl don't get the attention they deserve in our fast-paced world. Hunters know this. Children, too. My daughter's favourite animal is currently the cheetah, who, she solemnly informs me, must learn to sneak up on prey before attacking. Even the world's most famous sprinter relies on painstaking progress.

You could say I'm aging, but I prefer to see my devotion to slowness as a humble act of civil disobedience. Focusing my attention on, say, a two-millimetre-per-year rise in sea levels is my passive-aggressive, middle-class, Canadian white guy way of protesting the Great Acceleration of our times.

Because I don't know if you've noticed, but every minute of every day feels urgent. This is true even if you're mostly a stay-at-home dad whose life didn't change very much when lockdown began in 2020, since playing hide and seek with an introverted four-year-old pairs well with a Twitter addiction. Both the child and the platform enforce total immersion in the present moment.

That's not all bad. Books have been written about the importance of living in the moment. But enlarging the present to such an extent does have side effects. One I've been noticing lately is the erasure of past and future.

And what's the urgency, exactly? We all agree, even with our bitterest ideological opponents, that the world is in trouble, that things are breaking down faster than we can put them back together. But which things, and who is to blame, and what's to be done? Here the agreement disintegrates, even — especially — among allies, blown apart by a barrage of new and conflicting information. Democracy's collapsing in the moment of its triumph; the age of oil is ending while we burn more of it than ever; fifty species went extinct while we enacted laws to protect them.

When things move this fast, velocity starts feeling like the enemy. But maybe I'm wrong to think so. Maybe I should consider what velocity enables: a plane to lift, a boat to hydroplane, a bicycle to hold itself upright. Civilization, too, needs forward momentum. Another word for that is progress, though this implies direction, which gets us back into the realm of bitter disagreement as to where exactly we should go. Instead of settling on a route, we've wound up accelerating in many directions at once, at a rate that is itself accelerating, which is an unnerving curve to be on. At the speeds we're approaching, emergent properties begin to appear. Flight and combustion come to mind.

Thrilling, sure. Terrifying, too. But clarifying? I never used to think so.

↞〰↠

The one time in my life that writing felt quick and easy was also the most harrowing: the immediate aftermath of Kenya's 2007 federal election, which brought the country to the brink of civil war. Not many people outside Kenya remember it today, but for a few terrible weeks the world was transfixed by the stories and images that hundreds of journalists from around the world filed from East Africa's so-called oasis of peace. I was one of those journalists; the election happened to fall in the middle of the year that I spent living in Nairobi.

The story followed a familiar script. Election day itself went fine. There were no major irregularities reported as polls closed and the vote counters got to work under the watchful eyes of monitors. By midnight, the opposition leader had a commanding million-vote lead, and Kenya went to bed thinking it was over. We awoke to news that an improbable number of election officials all over the country had turned off their phones; these were the people who reported the official counts for their electoral districts, and nobody could get a hold of them. When they started flickering back into contact, several offered new vote tallies that varied greatly from the unofficial counts they themselves had shared on election night. As the new numbers dribbled in, the opposition leader's unassailable lead diminished, then disappeared, then turned into a 200,000-vote loss before the country's eyes. All the while, supporters of both presidential contenders started amassing in the streets. Soldiers gathered, too. It was like watching a river rise beneath a driving rain. Three days after the election, the country's chief electoral officer finally announced the "victor" on state television from behind a military cordon, and the river burst its banks.

Kibera, the city's largest slum and an opposition stronghold, was on fire before the victory broadcast ended. I was working for Kenya's *Daily Nation* newspaper, and my editors sent me to Kibera with a Kenyan photographer who'd grown up there. We drove to a small hill on the edge of the city that overlooked Kibera, where several television crews were already gathered. For a few indecisive minutes, dozens of journalists from Reuters, the BBC, Al Jazeera, and other outlets contemplated the smouldering sea of tin roofs spread out below us. Flames were starting to lick up and meet the molten sun dropping into the horizon, while a steady stream of grim-faced young men flowed past us to join their neighbourhood's mass protest. One by one, the camera crews pulled away for safer locations, but Moja, the photographer I'd come with, was unfazed. He urged me to join him. The sun dropped out of sight as I followed him into the labyrinth, and by the time I'd decided I'd made a bad decision, it was too late to change course. Night had fallen, and I was lost.

All I could do was stay glued to Moja's side while he strode through the maze, calmly snapping photos. Manic groups of people accosted us every ten steps, demanding we take their picture and tell the world what just happened. Many were teenagers, most clutched a machete or a stick or a bottle. They grabbed us, ordered us to drink with them, veered from laughter to fury and back again, teasing until they weren't, angrily enjoying this moment of collective release. We would pull free of one group with a combination of joking, pleading, and earnest promises, then march into the next. Every so often I lost sight of Moja, panicked until I saw his red shirt again and swam toward it. Shops were being looted; an electrical box exploded overhead; gunfire popped in the distance, coming nearer as the military — or maybe just armed supporters of the president — pushed in. We waded in the other direction, no longer trying to conduct interviews, focused solely on getting out. At length — an hour, three hours later? — we entered a section of Kibera where all was dark and quiet and wandered out to safety.

The next morning, as editors and reporters gathered in the newsroom, everyone thought the violence was concentrated in Nairobi. By noon we realized every other city in the country had gone through the same thing. Over the days that followed, reports started coming in from the countryside: Militias were blocking highways; whole villages were attacking one another; churches were packed with stricken families and surrounded by aggressors.

One of those churches was set on fire with four hundred people inside; in the stampede to escape, twenty-five children were trampled or burned alive.

My white skin conferred a sort of diplomatic immunity that most Kenyan journalists didn't enjoy. So, I travelled. I found myself entering other conflagrations throughout the countryside; none so intense as that night in Kibera, but many that left dead and wounded and homeless people in their wake. I learned to recognize the scent of a charred cadaver. I ran from tear gas and water cannons, skipped rope with small girls in displacement camps. The articles I wrote during that period were remarkable for no reason other than how effortlessly they emerged. The pace and magnitude of everything I witnessed conferred a sustained, vivid clarity I'd never known before and haven't since.

By the time Kofi Annan negotiated the power-sharing agreement that pulled Kenya back from the brink, I'd borne intimate witness to a cataclysm in which over 1,600 people were murdered and another 600,000 chased from their homes. It wasn't civil war, but almost. Close enough for me to grasp the raw energy that war correspondents imbibe — I slept four hours a night for a month without ever feeling tired — and the risk they run in doing so. Not of death or injury (though there's that) but, perhaps worse, the loss of empathy, of growing the "spiritual calluses" that the American journalist Garry Wills once wrote of.

Don't get me wrong. So long as there is war, we'll need people to bear witness and describe, decipher and explain it. Not all those who do grow spiritual calluses. But it's a very particular calling, and it isn't mine.

My brush with it was instructive, though. We all know disaster is riveting. But not till I experienced it, rather than watched or read about it, did I understand *how* disaster amplifies reality, how it grabs hold of our senses and our psyche, turns knowledge into feeling. That grip on our attention, I learned in Kenya, is largely a function of velocity. All the terrible things that happened after the election had been happening before it, too, just much more slowly. People in power had been killing their opponents, and buying votes, and stealing money, and abandoning children to scavenge on garbage heaps, for decades. A country cobbled arbitrarily and violently together by an occupying British army less than sixty years ago had always been something of an illusion. But nobody was shocked until the process sped up. Only when the slow violence of corruption and crushing inequality and simmering land feuds accelerated dramatically did everyone recognize the calamity it had been all along. Only then did it seem real.

Even a minor catastrophe — a fender bender, say — will absorb us completely if it happens fast enough. But let it approach at a crawl and we humans can ignore a disaster as big as the sky.

↞∿↠

It is early January when we start hearing about a strange new virus in Wuhan. It was first announced the day before New Year's Eve, but there is a lot of competition for things to worry about, and this novel coronavirus still feels like a distant threat. One of my friends, a science writer, is the only person I know who seems worried. She's been monitoring its progress, describes Wuhan as a bonfire that is throwing sparks into a parched forest. I'm not nearly as concerned as she is. I think back on SARS, Ebola, H1N1; they all came and left without touching my life. So will this.

On January 20, the first known spark lands in North America (of course, others have already landed, undetected) — a thirty-five-year-old man in Washington State's Snohomish County, two hundred kilometres south of my bedroom. Two days later, Johns Hopkins University launches a website that tracks the spread of COVID-19 around the world. Now politicians, researchers, health authorities, and the general public can watch the virus spread through 188 countries "in real time." The dashboard includes a graph showing the daily global total of new infections; the graph's slope will soon resemble the western flank of Mount Olympus.

But outside of China, the numbers remain low. Canada doesn't have a single case yet. The first impeachment of the forty-fifth president of the United States is underway and dominating news cycles. The verdict is a foregone conclusion, but the testimony is spell-binding. I forget about everything else. An American president is caught extorting a foreign leader for political dirt on his opponent, and half the nation shrugs.

On February 19, a Champions League match in Milan draws forty thousand fans. The phrase "super-spreader event" is about to enter the global lexicon. Over here in North America, it still feels like someone else's problem. At the end of February, my wife and I fly to Hawaii for a family reunion and try to prevent our daughter from touching too many things at the airport.

We come home ten days later. Italy, Spain, South Korea, and Iran are all in flames. Italy's hospitals "are like the trenches of a war," according to the wife of a man infected at that soccer match in Milan. Embers are now smouldering on every continent. Canada records its first COVID-19 fatality on March 9. On March 11, the World Health Organization declares the epidemic is now a pandemic — the storm has intensified into a hurricane. The next day, our prime minister's wife tests positive. This is the moment it hits home for me. Justin Trudeau goes into quarantine and begins a ritual of daily briefings from his front yard. On March 18, Canada formally locks down: schools close, restaurants close, sports stadiums close, the border closes to everyone but our own. "Let me be clear," Trudeau warns the three million Canadians still travelling. "If you're abroad, it's time to come home."

All this before a thousand Canadians have tested positive. That's less than 0.003 percent of the population. We're lucky: lucky that other countries got hit first, lucky to have a public health-care system, lucky our politicians heed doctors' advice, lucky we still have enough trust in our elected leaders to do what they say when it matters. Lucky, perhaps, that Sophie Grégoire got sick and jolted us awake. We aren't waiting for bodies to start piling up. We aren't talking about COVID-19 as *over there* or *coming soon*. It feels here and now and everywhere, even though it isn't, yet.

April 1, and the entire global economy has ground to a halt. Boris Johnson is the first world leader to fall sick; he winds up in a hospital hooked up to an oxygen tank, putting an end to Great Britain's pursuit of herd immunity. The Queen is roused to speak. The Great Acceleration hits the brakes. Every major city in the world feels like that scene in *Vanilla Sky* where Tom Cruise walks through an empty Times Square in the middle of the day. It's not quite that crazy — the streets aren't *totally* vacated — but traffic has disappeared for the first time in our lives. Care homes for the elderly become death traps. The virus infects our very language: "Before Times," "Zoom fatigue," "social distancing." My daughter calls this new reality "the coldness," because we have explained to her that there is a bad cold going around. A dedicated introvert, she is delighted by this turn of events. We joke among friends that she's been training for this moment all her life. I can't imagine what having more kids, or older kids, or a home smaller than our 1,200-square-foot townhouse would be like. One day in mid-April the city cordons off all the playgrounds with yellow tape, like crime scenes. Ada nods stoically and learns to ride a bike. My abiding memory of the spring of 2020 becomes jogging beside my daughter on the sidewalk while she pedals madly, laughs with rare abandon, doesn't even try to use the brakes.

↞∿↠

A thing that I like about spending time with a young child is the way some of their potential energy rubs off on you. For a three- or four-year-old, anything at all seems possible. The miracle of existence is so fresh and strange, the rules of physics and biology still so unknown, that they perceive no limits as to what might happen next. Birds can fly, so why not us? Humans can talk, so why wouldn't dogs? Why shouldn't it be possible for a person to eat this little piece of cake and grow tall enough to step over that building? It isn't just the good fantasies that seem possible but the terrifying ones, too. The way she whimpers and clutches me when the cartoon monster appears, I know the threat of that fanged beast jumping through the screen and eating her is as tangible as an actual rabid dog would be, barking in the living room. It's not a very adaptive trait, but it is instructive. Children are constantly searching for possibility. As we get older and lose the protections of childhood, we must become aware of limitations if we wish to survive. Gravity, for instance. But most of us, certainly myself, also take it too far. We become settled in our ways, mistake some of this universe's habits for rules.

The capacity to grasp what's possible, combined with the ability to make it happen, is a trait common to artists and leaders, innovators of all stripes. It's a neutral quality — Martin Luther King had it, but so did Hitler. Bad things that most of us can't imagine are possible, too.

As 2020 gathers speed, that childhood sense that anything could happen settles over the adult world. Everything — our politics, our economies, the most intimate aspects of our daily lives — is thrown into flux. Potentiality permeates each new day with threat and inspiration. The person you brush into at the grocery store could infect and kill you; your job could disappear; your company could get rich; everyone on Earth could stop flying; the United States could become a totalitarian death cult or flower into the most progressive government in history. Or anything in between.

↞↝→

While my daughter races down the sidewalk at approximately eight kilometres per hour, civilization squeezes its own brakes. The pandemic has accomplished what no other portent of death or doom could. We can't know exactly how much carbon we're saving, since, as the journal *Nature* notes, "systems are not in place to monitor global emissions in real time," but proxies are available. It looks like turning off most of the world's internal combustion engines is enough, at the height of the lockdown, to bring civilization's daily exhalation of carbon dioxide down by 17 percent.

Is this good? Is it bad? It is both. Emissions have been going resolutely up for over a century now. There was a brief pause during the last financial meltdown in 2008, but the subsequent acceleration more than made up for it. The explosion of renewable energy over the past decade has done nothing to bend that curve; in 2018, bitcoin mining alone consumed as much energy as all the world's solar panels produced. So if anyone had suggested in 2019 that emissions would drop 17 percent within a year, the environmental community would have done cartwheels.

But the price tag is outrageous. It's exactly as expensive as the fossil fuel industry always argues it would be to put them out of business. Millions are out of work, and every government on Earth is racking up debt levels not seen since World War II — on top of a death toll with no end in sight. People are trapped in their homes. The world's marginalized communities are terribly overrepresented among the dead, ill, and unemployed. Anyone singing about lowered emissions from the rooftops is well beyond tone-deaf. Short-sighted, too. By the end of the year, global carbon emissions will be back to pre-pandemic levels, reducing 2020's cumulative drop to roughly 6 percent.

Rob Jackson, an environmental scientist at California's Stanford University and chair of the Global Carbon Project, puts it this way: "We don't want a Great Depression to be the reason for our carbon emissions drop."

Writing in the *Guardian*, Australian journalist Jeff Sparrow describes the "awful paradox" that COVID-19 has revealed: "Capitalism must expand or lapse into crisis. But an economy dependent on perpetual growth must, at some stage, come into conflict with the limits of the natural world." The only way to save society, it seems, is to ruin all our lives.

Plagues may be older than the Bible, but they didn't used to come around so often. A 2008 study published in *Nature* found that ecological destruction has sped up the rise of new infectious diseases by a factor of four in less than a century. The study identified 335 new diseases between 1940 and 2004, two-thirds of them zoonotic — they passed to humans from animals — including SARS, Ebola, and HIV. The authors wrote that almost half of these new diseases "arose from changes in land use," which is a nice way of saying chopping down forests.

Nature is ramping up the kind of signals our species notices: forest fires, floods, droughts, hurricanes, and yet another virus. The slow roll of environmental collapse is gathering speed. It is entering real time.

I'm now hearing the expression "real time" every day. Every time someone says it, I'm reminded that our electric sensitivity to immediate physical threats is the reason our species is here today. That's a good thing. But real time also exposes a corresponding blind spot; it implies that slower-moving processes are somehow less than real. Tectonic plates don't shift in real time but in geologic time. Ecosystems develop over the course of deep time; they unravel much faster but still too slowly to perceive with the naked eye — *not* real time. Except for the people who study such things (and so tend to see the world through time-lapse goggles), glaciers do not melt in real time. Forests don't disappear in real time. Coral reefs don't die in real time. As a result, we fail to treat these things as emergencies. Never mind that unchecked climate change will kill millions more humans than COVID-19 — very few of them will die today.

↞↝

There are other key differences between this pandemic and the global ecological collapse it palely foreshadows. Perhaps the biggest is that politicians can still credibly claim this crisis shall pass — that all we have to do is buckle down and weather the storm, and then everything can go back to the way it was.

In half a year, when the second wave takes hold, Canada's finance minister, Chrystia Freeland, gives a speech articulating this very point of view. "The economic shock that we're witnessing as a result of COVID-19," she says, "is not due to a design flaw in our economy or in our businesses. We didn't get here because of greed or recklessness. The market isn't correcting for a flaw. This was a completely exogenous shock. Our citizens and our companies are suffering through no fault of their own."

That's mostly true — but the shocks coming after this pandemic, the ones that will come from climate change and environmental collapse, are in fact due to a design flaw in our economy and in our businesses. Greed and recklessness are very much part of that equation. Small wonder the only climate Freeland mentions in her speech is the economic climate. Both reassuring and terrifying, then, to hear her say that her overriding goal is to "ensure our economy comes roaring back, stronger than before," as soon as the virus is gone.

Back to normal. But normal, as writers and activists across the land are shouting in all-caps, is what got us here. Normal is the problem.

Already in April, with the first wave still building, that much is obvious. Even during this totally miraculous, unprecedented, and fleeting 17 percent drop in daily carbon emissions, we're still dumping as much carbon dioxide into the atmosphere as we were in 2006. That was the year *An Inconvenient Truth* came out. It was not a year of sustainable carbon emissions.

Apparently getting everyone to stop driving and flying isn't enough to save the world. That's because "most of the momentum destroying our Earth is hardwired into the systems that run it," Bill McKibben writes in the *New Yorker*. For 2020 to be more than a blip, we'll have to "rip out the fossil-fueled guts" of those systems. That means many things, from industrial agriculture to home heating, but above all it means converting the world's fossil-fuel-based power plants into renewable energy grids.

But this is all fancy talk. Before we go ripping any guts out, we have to put some bandages on. The patient is bleeding to death. You can't ignore a symptom, or theorize it into submission, if it's killing you. Can you?

People try. As April turns the corner into May, it comes as no surprise that the world leaders who rose to power in part by denying, or ignoring, or minimizing climate change are now employing the same strategy against COVID-19. In Russia, Brazil, Australia, India, and, of course, the USA, the result is always the same: volatility.

Stock markets set record highs and lows from one day to the next. The price of oil falls all the way to nothing, then keeps going; by the end of April, West Texas crude is trading at negative $40 per barrel, an incomprehensible result of oversupply and collapsed demand. Tell that to my peak-oil-obsessed twenty-five-year-old self. Remind him, before he celebrates how wrong he was, that this isn't necessarily good news for renewables. Cheap oil is hard to compete with. The Canadian government soon issues a report describing COVID as a "potential extinction-level event" for clean energy. Yet suddenly we learn that renewables provided more power to the U.S. electric grid than coal for every single day of April. In the U.K., solar power helped set a new record: eighteen consecutive days in which no coal-fired power plants were needed. The International Energy Agency announces that solar power has surpassed natural gas and coal to become the cheapest form of electricity on Earth. But only in optimal conditions. Not in Alberta, which opens fifty thousand square kilometres of the Rocky Mountains to coal mining.

On May 25, George Floyd is murdered. A police officer, Derek Chauvin, kneels on his neck for eight minutes and forty-six seconds while Floyd calls for his mother, then dies. America rises in his place. Over the following two weeks, over two thousand protests are held across the country.

It's impossible to say exactly how many people turn out to march, but informed estimates range from fifteen to twenty-six million; even at the low end, that's orders of magnitude more than the civil rights marches of the 1960s. Back then, almost all the protesters were Black. Today, everyone joins in. The Pew Research Center finds that Black Americans comprise 17 percent of the protesters, Hispanic Americans 22 percent, and Asian Americans 8 percent; 46 percent are white. According to a comprehensive analysis by a non-profit called the Armed Conflict Location and Event Data Project, 93 percent of the protests are peaceful. Support for Black Lives Matter jumps from 67 percent to 78 percent of Americans, rising as much in two weeks as it did in the previous two years.

Yet support for Donald Trump doesn't diminish; it hardens into militancy. The national police union endorses his bid for re-election. As with climate change and COVID, racism is both denied and given licence to flourish.

The notion that any one phenomenon can be understood in isolation, that things like racism and inequality belong to a separate conversation from ecological collapse . . . ha. The Great Acceleration has achieved the speed of flight. It's become the Great Convergence. Those millions of demonstrators marching in the streets marked the end of our global pause and the isolation that defined it. Something new is being born, some glittery fanged creature emerging from a cocoon and stretching still-wet wings.

"What we are witnessing right now is the opening up of imaginations," says Opal Tometi, one of the founders of Black Lives Matter, in an interview with the *New Yorker*. "People are really trying to show up in this moment for black people, but I think they are also doing it because they have been mad for a minute, almost like this pandemic was a pause, and they were able to think about what would justice look like, and what is actually going on, and they have been able to reflect on what is going on. I think they have been not O.K. for so many years, and they are finally saying, 'Hey, we are going to take it to the streets and say we are going to show up in solidarity with you.'"

Two days later, on June 5, Trump calls in the army to clear peaceful protesters from in front of the White House so that he can be filmed walking from the Oval Office to St. John's Episcopal Church. He doesn't enter the church or offer a prayer, but instead brandishes a Bible at the cameras in the manner of an infomercial huckster, offering viewers two revelations for the price of one.

Almost immediately afterwards, America's top military commander, General Mark Milley, chairman of the Joint Chiefs of Staff, apologizes for his role in the operation. "I should not have been there," he says. "It was a mistake." Instead of leaving it there, he adds that he, too, was outraged by the killing of George Floyd and sees the current protests as a response to "centuries of injustice toward African Americans." The anti-Trump world breathes a sigh of relief. We have been imagining what role the army would play in the inevitable election chaos for three and a half years now.

Time keeps compressing. Historical processes that normally span a decade or a generation or a century unfold over the course of weeks. Days. Hours. Eight minutes and forty-six seconds.

July offers the slightest of reprieves. Most of the world eases out of lockdown; life resumes a semblance of normality. Time slows down a little. We meet friends for picnics at the park; playgrounds are open again. Our daughter surprises us with a new confidence, as though emerging from her own cocoon. She learns to tell left from right, though not yet politically. She understands secrets and surprises, begins to rhyme, starts grappling with what makes a joke a joke.

This is a question many of us have been asking as we watch the leader of the free world make one outrageous statement after another. He suggests COVID-19 can be treated with bleach, jokes that he wants eight more years in office. Should we take him literally? Or seriously? Or both, or neither? Everything he says is so absurd, it should be comical. The news media, for the most part, is as confused as we are. Those journalists who have seen other countries travel down this road recognize what's happening, but it falls primarily to historians and late-night talk show hosts to translate the lunacy emanating from the White House.

America seems to have held the virus at bay through the sheer force of magical thinking. But the spell breaks in the middle of our July picnic. A thousand Americans start dying of COVID per day, a number that seems shocking. Soon it will double, then double again. The magical thinking takes this as a call to action. On July 8, Vice-President Pence says there are "early indications" the curve is flattening in Texas, Arizona, and Florida, the worst hit states at that moment; three days later, Florida becomes "the new Wuhan" by surpassing fifteen thousand cases in a day. Wearing a mask becomes a sign of political affiliation. Trump's term in office, which began with a promise to ban Muslims and

Mexicans, is closing with American tourists banned from all but eight of the world's countries.

"We are watching history unfold in real time," writes Charlie Warzel, a media and technology columnist for the *New York Times*. "For years now, no matter where you live, the myriad horrors of the world have been just a convenient swipe or tap away on our phones . . . But the experience, no matter how chaotic or hostile, has always felt siloed — contained to a specific city and particular accounts. Tweets about clashes between far-right protesters and anti-fascist protesters would collide in my feed alongside live tweets of a sporting event or dispatches from a campaign trail." It's fitting that Warzel describes this moment through the lens of Twitter, the ultimate real-time venue. "It's difficult to describe any online experience as universal," he writes, "but the scale of these protests feels all-encompassing in a way that I've never experienced: a singular, collective trauma dominating our national consciousness. There are no other channels to watch, no distractions."

The *New York Times* reports that the Trump administration has taken advantage of the pandemic's grab on our attention to roll back one hundred environmental protections. Joe Biden drops a $2-trillion climate plan to decarbonize the country by 2035. The Great Lakes are warmer than they've ever been. The oil majors are reporting second-quarter losses too enormous to comprehend: $1 billion for Exxon, $8 billion for Chevron, $17 billion for BP. Alberta's oil and gas revenues for 2020 drop from the pre-pandemic forecast of $20 billion to $600 million. Across the U.S., the oil and gas industry sheds 100,000 jobs. Collectively, in the first half of 2020, the G20 spends $200 billion on fossil fuel bailouts and barely half that on clean energy.

In August, climate change season begins. The Canadian Ice Service announces on the first day of the month that the largest ice shelf in the Canadian Arctic has collapsed virtually overnight: the Milne Ice Shelf, covering eighty square kilometres on the edge of Ellesmere Island, lost 40 percent of its surface in a few days at the end of July. On August 13, researchers report that Greenland's ice sheets have passed the tipping point — meltwater is draining faster than snowfall can replace it, so that even if global warming stopped today Greenland's glaciers would continue to disappear. "Greenland is no longer changing in geological time," they write. "It is changing in human time." On August 16, an electrical storm strikes California and unleashes eleven thousand bolts of lightning without a drop of rain, kicking off the worst fire season in the state's history. "If you are in denial about climate change," says Governor Newsom, "come to California." Five of the state's biggest fires in history burn this year, including the first million-acre blaze. The fires spread to Oregon, then Washington, blanketing the entire West Coast in record-setting smoke. The smoke blows east, mixes with Hurricane Laura, which slams into the refinery-laden coast of Texas and Louisiana as a Category 4 storm halfway through a Republican National Convention that is too depressing to describe.

The violence of those fires and hurricanes is matched joule for joule by the violence of America's coming election. Trump declares the only way he can lose is if the election is rigged, then hints that his followers should pre-emptively take their guns to the streets. The hint is taken. On August 25, a mother drives her armed seventeen-year-old son across state lines to Kenosha, Wisconsin, so he can fight Black Lives Matter protesters. Moments before he murders two of them and blows the arm off a third, he is filmed receiving water and praise from the police. He becomes a patriotic hero to Republicans. "How shocked are we that seventeen-year-olds with rifles decided they had to maintain order when no one else would?" declaims Tucker Carlson.

Amid all this, as August draws to a close, Jeff Bezos becomes the first person to amass $200 billion, Apple's market share surpasses

$2 trillion, and almost twenty million Americans go on unemployment relief. Four times as many women lose their jobs as men. A Bloomberg News headline notes, "Some of the biggest money managers are vexed by the same paradox troubling everyone else: U.S. stocks are near an all-time high, but the world still seems to be falling apart." Exxon falls off the Dow Jones Industrial Average after a ninety-two-year run.

September, and America enters its election home stretch. It feels like the country is preparing for war. Pundits start discussing a nightmare scenario in which Donald Trump edges out a slim lead on election night based on in-person ballots, but then mail-in ballots start coming in and put Biden over the top. Ada begins kindergarten.

↞↝

There is, of course, another word for what I've been describing. That word is "intersectionality." Our identities, our problems, our solutions are multifarious and all bound up together. It's not original to say so, but people keep forgetting. Now 2020 is reminding us at the top of its lungs that everything is connected. The world has gone under a microscope. Anyone can look through the glass to see where the fault lines intersect and submerge, converge and buckle. Climate change and racism, denial and corruption, activism and democracy, war and peace. There they all are, squirming on the petri dish, caught in the act.

Senator Ed Markey, co-sponsor of the Green New Deal, puts it this way after winning his own primary: "There will be no peace, no justice, and no prosperity unless we stop the march to climate destruction."

By then, fires are peaking across the entire West Coast. San Francisco is dark at noon, Portland's mayor declares a state of emergency, Los Angeles County records its hottest day in history.

The fires melt infrastructure and sewer pipes, pouring benzene and other toxic chemicals into the water supply. On September 14, Donald Trump finally visits California and says, "It'll start getting cooler, you just watch."

Now things really start to spin. At a UN General Assembly meeting on September 22, China's president, Xi Jinping, announces that China will be completely decarbonized by 2060. This is a huge deal — China produces as much carbon dioxide as the U.S. and Europe combined. The next day, Governor Newsom announces that California will ban the sale of gas-powered cars in 2035. A week later, the world's largest wind and solar power generator, NextEra Energy, overtakes Exxon, the world's largest oil company, for market capitalization. Boris Johnson pledges to make every home in Britain run on wind power by 2030. Already this year, the European Union will generate more power from renewables than fossil fuels for the first time in history.

But scientists warn that 40 percent of the Amazon is now poised to turn irrevocably from rainforest into savannah — less rain, more fire. On October 6, Hurricane Delta becomes the fastest storm ever to strengthen into a Category 4 in the recorded history of the Atlantic. Swiss Re, one of the world's largest reinsurers, reports that a fifth of the world's nations are at imminent risk of total ecological collapse. This, it notes, has financial implications. Climate-related disasters will cost the United States alone $95 billion in damages this year, which ties 2016 for the hottest year in recorded history.

On the eve of the election, more than ninety-five million Americans have already voted. Over 230,000 have died of COVID. Eric Holthaus, an American climate journalist, writes: "Decades from now, when we write the history of 2020, the most important and most lasting trend will be that — seemingly all at once — people's voices mattered. Poor people and people of color, young people, marginalized people from all over the world, rose up in a too-late-but-right-on-time revolutionary movement to elevate the joint cause of racial justice, climate justice, economic justice into a broader Movement of Justice . . . The racial reckoning, the climate emergency, and the crushing consequences of the pandemic — they're all coming to the fore at the same time. Let these truths radicalize you. And then, get to work and re-write your future."

On November 3, America starts counting votes. Each day of the next ten weeks is its own novel. I think about Kenya a lot.

↞↝→

On January 8, 2021, the *Daily Nation*, where I'd been working thirteen Christmases ago in Nairobi, dedicates its front page to one of the many iconic photographs that emerged from the U.S. Capitol two days prior. The image of a smirking rioter sitting in Nancy Pelosi's office, legs splayed open with one boot on her desk and an American flag tossed across her cabinet, has this headline below it: "Who's the banana republic now?"

Fiona Hill, the former National Security Council official whose testimony had been a major part of Trump's first impeachment trial, explains five days later why the thing we all just witnessed was in fact "a slow-motion, in-plain-sight attempt at self-coup." Writing in *Politico*, Hill notes that many commentators don't think the events of January 6 qualify as an attempted coup. They prefer instead to call it an insurrection, or a riot, or an uprising: "These observations are based on the idea that a coup is a sudden, violent seizure of power involving clandestine plots and military takeovers." But Trump did all those things, Hill notes; he just didn't do them suddenly.

He declared "election fraud" within hours of the polls closing, three days before the votes were fully counted. He hurled over sixty lawsuits at courts he'd spent four years packing. He fired Mark Esper, secretary of defense, along with several top officials at the Pentagon, and replaced them all with loyalists, around the same time he began planning his "stop the steal" rally in the Capitol. And then he sent his armed supporters in to "fight."

Even so, people still don't like the word "coup," partly because Trump's mob was so dishevelled but also, Hill concludes, because "his actions were taken over a period of months and in slow motion." As a result, we came within moments of seeing a substantial portion of the American government assassinated on Instagram Live.

"The culmination," Don Lemon says on CNN, a few hours after the mob went home, "of five years of lies, playing out right now, in real time."

Something essential about those years has just been irreversibly exposed. Time keeps hurtling by after the curtain falls away; the aftermath merges with preparations for the inauguration, which merge with reports of further attacks being planned for every state capitol in the nation. But an inflection point has been reached. Events begin to slow, almost imperceptibly. We've entered a new curve.

Trump is finished. Soon the pandemic will be, too, thanks to the fastest vaccine creation in history. Two fevers broken, though both still have damage to inflict. Both could still mutate into something even worse. The most heavily armed nation on Earth is more fractured than ever since its civil war, while nearly four thousand Americans a day are dying of COVID-19. A similar dynamic is in play for climate change, racism, and inequality, solutions to all of which are in the offing even as their consequences intensify.

Can the Biden administration inaugurate a period of healing? Will the world change course in time? Or have we simply triggered a new cycle of reactionary politics, a time of violent upheaval to be set against a backdrop of runaway climate change and ecological collapse?

Will society take flight or combust? Anything is possible.

THE SUSPENSION OF DISBELIEF

WE ARE BOMBARDED BY NEWS. It hails down from above and boils up from below, more news than there's ever been in the history of news. You'd think all this information would help us make sense of the world, but instead the opposite has happened. There's too much to know, too many parts moving in too many directions. Stories that ignore each other one moment contradict each other the next.

"Collapse of civilization is the most likely outcome," reads a typical headline in my feed, followed by another, just as typical: "2021 could be a boom year for stocks." You can play this point-counterpoint game with any subject. The protests are peaceful — see how they loot! We can't go further into debt — there's never been a better time to borrow! Amid the deluge, the presumption that any two people who watch the same thing will see the same thing has evaporated. That's an old story — which is to say, *hardly news* — but we've reached new heights in this accelerated age, and we're now at the point where observable facts like the size of a crowd or the temperature of an ocean are considered opinions, subject to passionate debate. There are as many realities today as there are smart phones. That's new, and therefore news.

In this environment, the very meaning of a word is up for grabs. "Antifa" might describe someone opposed to fascism (whatever *that* is), or a person who sets buildings on fire for fun. "Defund the police" could be a slogan for abolishing police entirely, or for spending more money on social programs. It's entirely up to you. Lamenting this trend and its implications for democracy, Barack Obama once famously insisted that "words *mean* something." But not even the power of the presidency could make it so. His successor had better luck pushing language in the opposite direction, and such is that man's legacy that words remain more vulnerable than ever. They can mean nothing, or anything at all.

It turns out whole segments of society can also be made to believe in nothing or anything at all. For example, at the height of Oregon's historic forest fires in September 2020, police emergency lines were overloaded. The mayor of Portland had declared a state of emergency, and communities throughout the state were being evacuated, but the panicked calls flooding police lines weren't about escape routes. People were calling in to report that antifa was lighting the fires.

↞↝

I once made the terrible mistake of trying to write a novel about this kind of thing. My protagonist was a radical environmental activist who couldn't tell if the world was going crazy or he was. How could it be, he wondered, that more people weren't worried about the destruction of the planet? Everyone was having so much fun. In order to make his confusion more believable, I made him a septuagenarian with a touch of dementia. Then one day, about two hundred pages in, Donald Trump got elected.

One thing you discover when you think long and hard about collective insanity is that it has more to do with deception than genuine madness. Individuals can lose their minds but — except in war — whole societies generally don't. What they do is fall for a lie, or many lies, then act accordingly. So when you call a large group of people "insane," what you're really saying is you haven't heard the lies they're being told; or if you have, you've yet to reflect on why people might believe them.

In the fall of 2016, I was living in the U.S., on a journalism fellowship at the University of Michigan in Ann Arbor. Officially, I was there to research the economic rationale for perpetual growth, the most popular lie I know. But that autumn's mega-merger of reality TV with cable news and social media wiped out every other waking thought. Michigan, a swing state that helped clinch Trump's victory, was at the heart of this terrible moment for the world. That made it a wonderful place for a journalist to be.

Not that Ann Arbor is Trumpland. It's a quintessential liberal college town, affably progressive and surrounded by conservative farmland with the odd abandoned factory. Drive through those rusting fields for half an hour and you reach Detroit, the engine of American growth in the twentieth century turned symbol of national decay, turned symbol of rebirth. In Detroit, you could walk from Trumpland to NeverTrumpland by turning a corner. But Ann Arbor never went through that kind of centre-of-the-universe whiplash. The placid streets of A2, as it's known, reveal an earnest town where nothing has ever gone explosively right or devastatingly wrong. Despite being a very white town in the middle of an industrial state, it is solid Democrat turf.

But even there, maybe especially there, you could feel the lies lurking beneath the surface. One manifested in the proportion of campus janitors and fast food workers who were Black. Another skulked in the parks and playgrounds that felt eerily quiet on even the sunniest weekday, because the most affluent country on Earth believes it can't afford parental leave.

A month into my stay in Ann Arbor, I learned that a decades-old industrial spill nearby had contaminated the groundwater, and this massive blob was seeping slow and steady toward the reservoir that supplies Ann Arbor's tap water. Everybody knew it, but nobody knew what to do about it, and so the general response was to talk about something else.

More disturbing still was the intellectual establishment's calm appraisal of Donald Trump's candidacy. The University of Michigan is a major centre of gravity for American public discourse, both an incubator and enforcer of the Way Things Are. So it was striking, in the seminars and lectures I attended throughout September and October, to see so many elite thinkers take it for granted that the vote would be close, because the vote is always close, because there are two parties in America and a symmetry of power is the antidote to one-party rule. If one of the two sides has chosen a braying mule to lead it, the professor's job is to extract segments of the braying and recombine them into phrases that resemble policy proposals, then discuss those proposals in a dignified manner. There's no cause to get upset about it, because this is how Things can stay the Way They Are.

For a less cerebral interpretation of events you only had to wander over to the Big House on the edge of campus, where the largest football stadium in the country offered live weekly clinics on American identity politics to 107,601 students at a time. My wife and I had never been to an American football game before, so one day we decided to check it out. Before the game began, a hundred thousand adults, plenty tipsy, held their hands over their hearts and belted out the American anthem; when the closing notes wound down, two fighter jets roared over our heads, close enough to smell the jet fuel. The hands turned into fists held high, pumping to a collective chant of "U-S-A! U-S-A!" Then the game began, providing an excuse for tens of thousands of people to keep chanting. Mostly they hurled schoolyard taunts in chorus. "Yooouuu SUCK!" was the one that rallied the most voices. That the insults started out as jokes didn't stop them from becoming heartfelt, spilling over into violence in the parking lot or among the tailgate parties that surrounded the stadium.

The strangest part of all this was how hypnotizing it was. Lectures and suits, warplanes and fists, cappuccinos served by the grandchildren of enslaved people — it wasn't till two in the morning on November 9, when the shockwave of Trump's victory knocked tears from our eyes, that I realized I'd fallen under the same spell as everyone else. I was surprised. We were all surprised, including, it was said, the new president himself. Nobody'd believed it could happen.

That a person like Donald Trump could become the leader of the free world said more about the "free world" than it did about Donald Trump. It also spoke volumes about the world's freest press, for it, too, was surprised.

That astonishment revealed the presence of a collective blind spot. Somewhere in our news-reporting solar system, a black hole was warping gravity. That black hole had spit out Donald Trump, a reality TV star who would henceforth be as blinding as the sun. There could be no ignoring such a sun once it took over the sky, but it remained necessary to squint and scan the margins, because any hope of removing it rested on locating the black hole from which it emerged.

And guess what? After four years of daily observation that ruined all our eyes, the black hole was discovered. It was hiding behind the lode star that has guided journalists for ages: neutrality. The foundational principle of our profession, the basis of our credibility as impartial referees, without which no game or political campaign can fairly proceed. Journalists are referees, and a referee must remain impartial. They can never pick a side.

This reverence for neutrality is easy to hack, especially when it's bound up in two-party symmetry. A core part of the referee's job is to call fouls, so if one team starts getting too many fouls, that team can claim the referee is biased. *Why are we getting all the calls? Why are you taking sides?* And the referee, ever mindful of the emotional crowd, maintains symmetry in the name of staying neutral.

When that happens, you get *Democrats say you should wear a mask, Republicans say no biggie, you decide.* You get *Republicans say climate change is a hoax, Democrats say it's real, you decide.* That's the hack. It hands a weapon to whichever side lies most by refusing to explain that one side is lying more than the other, or that one side represents 99 percent of the scientific community. It empowers bad-faith arguments by receiving them in good faith, and by giving them the same amount of attention as genuine arguments. It also gives the minor infractions of one team the same airtime as the blatantly criminal misdeeds of another.

This asymmetry didn't suddenly appear in 2016, and it didn't go away in 2021. It built up incrementally, masked by the slow velocity of its development and also by the fact that both sides do lie, and spin, and argue in bad faith. One side just kept doing it a little bit more, year after year, and kept being rewarded for it until the gap between them widened to the point that a man like Trump could take over the Republican Party. Suddenly, the truth was plain to see but terrible to describe. Suddenly, thousands of reporters in the United States faced a collective choice: describe the depths to which the Republican Party had fallen under their watch or stay neutral and stick to quotes.

The best, most searing reporting of American decline has always come from within the United States. Throughout Donald Trump's rise to power and over the course of his administration, countless investigative journalists worked without fear of sounding partisan to warn the nation of this strange new threat from within. Their relentless reporting was the only reason I knew about Trump's innumerable depravities, which you either know about already or won't believe if I repeat them now.

But wherever traditional journalists did manage to expose precisely what Trump and his corrupt coterie of loyalists were doing — from separating families at the border to tax fraud — the influence of social media rendered their work invisible to half the nation. Obviously, neutrality wasn't the issue here; far from being too cautious, the news spreading on the socials made a point of being reckless and utterly indifferent to the truth. As we discovered too late, unregulated social media platforms were particularly fertile ground for fake news, because fake news is more stimulating. It dispenses with the tedium of facts and evidentiary standards to go straight for the emotional jugular of wild accusation and tribal affirmation. It comes right out and says, *You suck.*

The intentionally blurred boundaries between social and traditional media, real and fake news, have given rise to what's now regarded as an epistemic crisis — a growing inability to distinguish fact from fiction. The general contours of this crisis are pretty well understood today, even if the solutions aren't. But in the early days of Trump's presidency the crisis was, if not exactly new, still dimly perceived by most journalists and the general public. All we knew was that it didn't seem to matter what Trump did or how relentlessly someone reported it; the story always disappeared in a torrent of symmetrical headlines and outrageous counter-accusations.

One person who saw this clearly at the time was Steve Bannon, Trump's close adviser during the campaign and one of the last to receive a presidential pardon (he'd been charged with fraud for pocketing donations meant to support Trump's border wall). "The real opposition is the media," Bannon famously said, "and the way to deal with them is to flood the zone with shit."

It worked. It's still working. If you aren't a news junkie, if you're a regular citizen with a job and family responsibilities and better things to do than spend hours each day parsing current affairs, what you behold in the newspapers and online and on TV is a tsunami of shit, a symmetry of shit splattered across the political spectrum, a kaleidoscope of shit hitting ten thousand fans all at once.

You can give in to the chaos, throw your hands up in the air, conclude everyone is lying and nothing will ever make sense. That's a popular response. It's also a form of self-deception, because the notion that you can't believe anyone is itself a belief.

I don't know if it's ironic or perfectly logical, but it's certainly true that people who say you can't believe anyone are often incredibly gullible. Long after Trump and Bannon were banished from the White House, the shit they and so many others sowed keeps fertilizing minds, allowing all sorts of laughable conspiracies to grow in the imaginations of tens of millions of otherwise rational souls. It's worst in America but hardly confined to it. Who doesn't have a few friends posting conspiracy theories on Facebook? Wherever there's an internet connection, you'll find a ready garden.

↞∿↠

Within a week of Trump's inauguration, his press secretary was talking about "alternative facts" and U.S. sales of *Nineteen Eighty-Four* were up almost 10,000 percent. A strong choice, frequently distilled on Twitter to a single quote: "The party told you not to believe the evidence of your eyes and ears. It was their final, most essential command."

In many ways, the Trump years did boil down to that line. But word for word, Orwell's essay "Politics and the English Language" is probably a better guide than *Nineteen Eighty-Four* for navigating the route public discourse has charted since the turn of this century. "If thought corrupts language, language can also corrupt thought," Orwell observed, as though predicting Twitter.

> The decline of a language must ultimately have political and economic causes: it is not due simply to the bad influence of this or that individual writer. But an effect can become a cause, reinforcing the original cause and producing the same effect in an intensified form, and so on indefinitely. A man may take to drink because he feels himself to be a failure, and then fail all the more completely because he drinks. It is rather the same thing that is happening to the English language.

Orwell died in 1950, the year after *Nineteen Eighty-Four* was published. While he was alive, he was better known for his journalism; it was his sharp observation of real life that made the fiction for which he's now famous so powerful. Not till after his death did writers learn to invert that formula and begin harnessing the techniques of fiction to fuel the way they wrote about real life. The New Journalism was founded by people like Truman Capote, Tom Wolfe, and Joan Didion, who learned to draw their scenes and develop their characters with painstaking detail. They generated suspense through hints and ticking time bombs, ensured their stories had a plot with an arc. Instead of making themselves an invisible fly on the wall, they cultivated a narrative voice. They wrote with style and swagger.

This marked a profound departure from traditional journalism. It was more engaging, because it deliberately played with our emotions, always seeking to surprise, or delight, or outrage the reader. Another difference was the way the New Journalism writers conveyed information: subjective impressions carried as much weight as objective fact. As in a novel, the way a person spoke could carry as much meaning as the words coming out of their mouth. That's how it is in real life, after all.

Ultimately, novelists and journalists have the same two overarching goals: Their stories must be both interesting and plausible. Those qualities are often in conflict (hence the need for technique), but thankfully for producers of fiction and non-fiction alike, humans are wired for story. We *want* to believe. We *want* to weep and laugh and cheer for characters real and imagined. It's a really remarkable thing, that dual state we enter when fully consumed by a literary work. It reminds me of the way my daughter can insist her favourite teddy bear really wants to swim in the tub with her, but if I say that Teddy must have gone for a walk since I can't find him, she'll cry, "That's impossible, *he's a stuffy!*"

Still, it's harder to create that state of mind in adults than it is in children. With age comes doubt; the childhood impulse to believe grows tempered by the skepticism of experience. That's a barrier all storytellers have to overcome. Not just novelists and journalists but screenwriters and advertisers, economists and environmentalists, politicians and profiteers, too.

It was a poet who coined the relevant phrase. More than two hundred years ago, Samuel Taylor Coleridge tasked himself with immersing his staid English audience in a supernatural tale about a sailor's harrowing voyage aboard a ship of ghosts and angels. The trick, he later wrote, was "to transfer from our inward nature a human interest and a semblance of truth sufficient to procure for these shadows of imagination that willing suspension of disbelief for the moment, which constitutes poetic faith."

How's that for a mouthful? Chock full of the stilted formulations that Orwell would have skewered for obscuring the point instead of illuminating it. You have to read it twice to make out what Coleridge means. But that dull-sounding experiment in suspending disbelief gave us *The Rime of the Ancient Mariner*, and it helped pave the way for a whole bunch of crazy characters thereafter.

Coleridge didn't invent the art of suspending disbelief; that's as old as storytelling itself. But by giving it a name he helped us understand and do it better, marking a singular contribution to an art that's come very far indeed over the last two centuries. This is mostly wonderful but also very dangerous. Because we've come full circle. First novelists mimicked poets; then journalists mimicked novelists; finally, in this golden age of conspiracy theory and unfettered social media, politicians and their propagandists are borrowing the tricks of journalism to suspend the public's disbelief from ever greater heights.

The result is a modern variant of something that Hannah Arendt described in 1951, a year after Orwell's death, just as the New Journalism was being born: "The ideal subject of totalitarian rule is not the convinced Nazi or the convinced Communist, but people for whom the distinction between fact and fiction (i.e., the reality of experience) and the distinction between true and false (i.e., the standards of thought) no longer exist." In other words, an epistemic crisis.

We were warned about this kind of thing. How did we not see it coming?

←〰→

It's hard to imagine, but as the 1990s were coming to a close, the free press and much of the world it informed were in general agreement that things were trending up. The Cold War had ended like a miracle, and Apartheid, too. In Latin America, an entire continent pried itself free of dictatorship. Huge swaths of Asia were emerging from crushing poverty, famine, and slaughter. Y2K came and went without a glitch. The internet was a shiny new crystal ball with limitless potential to solve the problems that remained.

A premonitory speed wobble arrived with the contested election of George W. Bush in 2000. Then came 9/11; then, infinitely worse, came the U.S. reaction to 9/11: a "war on terror" that spread terror around the world. Readers of a certain age will remember the moment Colin Powell, the U.S. secretary of state, presented the United Nations Security Council with grainy satellite images of long metal tubes stacked on the back of a truck. These were Saddam Hussein's weapons of mass destruction, Powell insisted. The mainstream press, from the New York Times on down, spread the lie unvarnished. The golden age of conspiracy theory was born.

That was how Alex Jones, founder of InfoWars, got his start. Long before he was calling the Sandy Hook massacre a lie or whipping up Trump's mob into a Capitol-storming frenzy, Jones was promoting a conspiracy theory born alongside the story of Iraq's weapons of mass destruction: 9/11 was an inside job. That conspiracy gained enormous traction among people who loathed and distrusted the Bush administration. Among those poor lost souls was me.

The theory that the Bush administration was involved in the destruction of the Twin Towers was very appealing to an aspiring journalist. The actual truth was just so implausible: the bumbling ineptitude of the hijackers; the fantastic coincidence that the U.S. air force was distracted by a training exercise that very morning; the way the Twin Towers collapsed so perfectly; the way a forty-seven-storey building all the way across the street from the Twin Towers collapsed hours later, pancaking down in precisely the manner of a controlled detonation without having been touched by either plane. Not only were the circumstances fishy, the motivations were clear as day. Now the U.S. could invade the Middle East and seize all the oil it liked, as the price surged past $100 per barrel and domestic production was dangerously low. In fact, the Bush administration had already drawn up plans to do just that, before 9/11.

The fact that the inside-job angle was so roundly (and rightly) scorned by respectable journalists only increased its appeal. These were the same people who were pushing for the U.S. to invade Iraq. I, for one, would never be so credulous.

I also didn't expect to do much about it. But in 2004, I took an internship at *Harper's Magazine*, headquartered in the heart of Manhattan at the marvellous address of 666 Broadway. The editor in chief was Lewis Lapham, a legendary journalist and historian who'd been at the helm of the second-oldest magazine in America since the year I was born. Lapham was in his sixties, a man from another era who wore tailored pin-striped suits, chose every word with gravelly precision, and carried himself with the patrician air of a slightly mischievous senator. His grandfather had been the mayor of San Francisco. He'd spent time with the Beatles in India. He chain-smoked in his office after New York banned indoor smoking, drank coffee all day long, hung around with the likes of Alec Baldwin, and told an excellent story about applying for the CIA that no one knew the truth of.

Like any senior editor, Lapham would occasionally recruit an intern for a random task you could never ask someone on the payroll to do. One day, he asked offhandedly if I was interested in looking into the inside-job angle. "I'm pretty sure it's full of shit," he said, "but you never know."

I spent every subsequent free moment in New York devouring whatever scrap of information I could find on the subject of 9/11. Imagine if I could be part of a *Harper's* exposé that made the nation shudder. The research mostly took place on the internet, though it also included a few meetings with fellow aspiring cloak-and-dagger types who looked over their shoulders a lot, and whose fates I ponder to this day. One of them was a paranoid but gruffly charming former cop named Michael Ruppert, who would soon attract a modest cult following for his book *Crossing the Rubicon*, which merged the 9/11 conspiracy with the Iraq War and peak oil. Over the following years, Ruppert grew increasingly irrelevant, bitter, and broke. If he could have held on a few more years, he might have found a new stride in the QAnon era, but he died by suicide in 2012, earning one last post-mortem dance in the conspiracy limelight when people took the note he left behind for evidence that he'd been murdered.

Long before that, I left New York and abandoned my brief, first, and only pursuit of a conspiracy theory. Instead of smoking guns, the clues of an inside job only ever led to more clues, and eventually the dearth of genuine expertise among those peddling the story grew too glaring to ignore. I still can't prove that the World Trade Center wasn't packed with explosives in advance of September 11, 2001, or that the Bush administration didn't divert the air force to clear space for the hijackers. But the experience wasn't a total loss. It taught me the core ingredients of a good conspiracy theory. For instance, the trick of zooming in on incongruent details, no matter how minor, and blowing them up as proof that an entire established narrative is incorrect (how could that flag be waving on the moon?). There is also the dressing up of armchair enthusiasts as "experts," their authority proven by degrees from universities you may or may not have heard of, in fields that may or may not be related to the subject they claim to be experts on. As those experts can attest, it's important to present your conspiracy theory in scientific language so that it conveys deep understanding. Finally, crucially, if you really want your conspiracy to go viral, make sure to ground it in an underlying cause that triggers a deep sense of identity — patriotism and religion, white supremacy and freedom all have good track records for bringing the masses on board.

All of which brings us to the heart of the matter: the denial of climate change.

"I was interested in your article today in the *Globe*," the email began. "As an environmental supporter with a science back-ground, it left me thinking about a couple questions."

So began a typical email I received in response to an article I wrote for the *Globe and Mail*, in which I compared the denial of climate change by certain world leaders to those same leaders' denial of COVID-19. The person emailing me was a retired engi-neer from Calgary (I looked him up). He understood science, and as far as I could tell, he didn't deny that COVID-19 was real and posed a genuine threat; all he denied was that the Earth's climate is warming due to the buildup of greenhouse gases.

"How will rational environmentalists deal with the emerg-ing science that shows warming isn't as bad as portrayed by the extreme activists?" he wanted to know. "Assuming you're rational and not an extremist? Even the IPCC scientists disagree with the alarmist position. I hope for Canada's sake we can be realists and not blindly follow emotional extremism."

The first thing to say about an email like this is that it was fairly polite. Being a white man spares me the abusive filth and death threats that others who write about climate change routinely receive. This never happens to me. What I get is thinly veiled con-tempt ("assuming you're rational").

But the thing that made this letter emblematic, the reason I quote it here, is the way it mixes absolute cynicism with total credu-lity. That's a defining characteristic of the conspiracy theory crowd. Huge swaths of the public now reflexively distrust genuine experts in every field, only to place their faith in transparent bullshit artists.

Not that there's no cause for distrust. It was experts, after all, who ushered in the nuclear policy of mutually assured destruc-tion. It was experts who promised for decades that smoking didn't cause cancer. It was experts who said we should eat meat and cheese and bread every day, and experts who reassured us that opioids were not addictive. There have been so many abuses of trust in this bewildering world that no one can be surprised the public has grown suspicious; it's where they choose to place their trust instead that consistently amazes, the equivalent of taking your savings out of a dodgy bank and handing them to a thief for safekeeping.

Now here we are, in an informational landscape crowded with inane conspiracy theories, believed or at least entertained in many cases by people who are genuine experts in other fields — that is, people who have demonstrated their capacity for critical thinking, yet believe that vaccines cause autism, or that Bill Gates created this novel coronavirus, or that the Democratic Party is a cabal of Satan-worshipping pedophiles. For sheer outlandishness and murderous intent, QAnon sits at the top of this pile. But in terms of size, global reach, political saturation, and pseudo-intellectual pedigree, QAnon remains an upstart in comparison to the grand pooh-bah of all conspiracy theories, which is climate change denial.

I don't just mean in terms of the number of people who believe climate change is a hoax, or pretend to out of vested interest. I mean the number of people who would have to be in on the conspiracy itself for it to be true. Even QAnon, for all its baroque imagination, limits the number of people involved in the conspiracy to approximately one-half of the American elite. That's penny-ante stuff.

For climate change to be a hoax, all those baby-blood-drinking Democrats would have to be in cahoots with tens of thousands of climate scientists from all over the planet, whose university department heads would have to be orchestrating the incubation of this hoax. They, in turn, would have to be recruiting an even greater number of journalists and politicians. This incredibly coordinated international coalition of conspirators would have to be cooking the books on temperature records from every corner of the planet, a project which would further require the participation of meteorologists from most of the countries in the world. The conspirators would also have to tamper with government-owned satellite imagery to make it look as though all the world's glaciers and both polar ice caps are melting, while thousands of biologists would have to pretend that the world's coral reefs are dying of heat. The Pentagon and NASA would have to be involved at the highest level. The National Science Academies of at least seventeen countries, and the global insurance industry, too. Just for starters.

The sheer scale of credulity required to believe climate change is a lie, combined with the scale of evidence that it is real, has lately forced many deniers to backpedal slightly. Canada's Conservative Party has written the playbook on this new form of climate denial. Instead of calling climate change a hoax, Conservatives here and abroad downplay its severity and emphasize unknowns. Climate change may be real, they nod, but it's nowhere near as urgent as people say. Much remains unknown. Economic collapse, on the other hand, is a clear and present danger that we understand very well. The cure can't be worse than the disease. So let's be rational.

That kind of talk acts as a shield for inaction. It also provides cover for the outright deniers, who still constitute a majority of Canada's official opposition — at their national convention in March 2021, the Conservative Party rejected a proposal to include the phrase "climate change is real" in their policy book. And that's what makes my Calgarian pen pal's email worth highlighting. Here was an intelligent, science-literate man who truly believes that climate change is an unmitigated hoax. He knows who he's voting for. If he was an isolated crank, I could ignore him, but he's not. There are millions of people like him, throughout North America and beyond.

One thing they all do is project their own errors onto us. When I read "I hope we can be realists and not blindly follow emotional extremism," I think of the emotional extremes that enable a grown man to hold on to the fantasy that there are no consequences to burning fossil fuel. I think of the term "fake news" and the people who deploy it. I think of the expression that preceded fake news — "junk science" — which was coined by Big Tobacco to describe studies that found a link between smoking and cancer. I think of a recent Fox News headline that screeched, "Democrats are living in a fantasy world of denial concocted by a complicit media." I think of Donald Trump accusing Democrats of trying to steal the election.

Above all, I think of Leon Festinger.

↞∿↠

Leon Festinger was the American social psychologist who coined the term "cognitive dissonance." In his 1957 treatise *A Theory of Cognitive Dissonance*, he made the deceptively simple observation that "the presence of dissonance leads to action to reduce it, just as, for example, the presence of hunger leads to action to reduce the hunger." Festinger was not the first person to notice this effect. But he was the first to systematically study it, pioneering a series of groundbreaking social experiments that revealed the things humans actually do to reduce the dissonance in their lives.

Festinger found two main categories to describe our most common reaction to dissonance. First, we can change our behaviour; second, if that's too difficult, we can filter the information we receive.

Take the example of someone addicted to smoking. There is a dissonance between the knowledge that smoking is harmful and the pleasure it provides. If a person wants to reduce that dissonance, one obvious option is to change their behaviour — quit smoking. But that eliminates the pleasure, too; so, another option is to filter their information. As Festinger writes, "The person might seek out and avidly read any material critical of the research which purported to show that smoking was bad for one's health. At the same time he would avoid reading material that praised this research." Such a person, Festinger adds, could also reduce the *importance* of the dissonance by reading about all the other dangerous risks in his life that have nothing to do with smoking, like driving and flying or, better yet, overeating, which smoking is known to inhibit. It's striking that he chose this example in the 1950s, just as the tobacco industry embarked on its three-decade campaign of dissonance reduction.

Few examples in real life are as straightforward as smoking, or as strictly personal. For one thing, no two cognitions exist in isolation; our knowledge and beliefs are bound up in an intricate web that adds up to our world view. World views may seem personal, but they're really not; they bind us to a community, crucial for a social animal like humans. World views are the fabric from which society is woven. They are personally held but also intensely political. And they are extremely sensitive to dissonance.

The greater the threat to our world view, Festinger found, the greater the extremes we'll go to in order to ignore conflicting evidence. That's why Galileo was forced to recant his observation that the Earth orbits the sun: if that had been acknowledged as true, then God had not made us the centre of the universe, which cast a central premise of Christianity in doubt, threatening the entire power structure of medieval Europe. Galileo was given the choice heretics have been offered through the ages: refute the evidence of his eyes — eliminate the dissonance — or die.

Not everyone refutes it as reluctantly as Galileo, who according to legend muttered, "And yet, it moves," right after he recanted (the seventeenth-century Italian version of "Psych!"). Reassessing deeply held beliefs is painful. Many of us would rather reject the evidence of our eyes and ears, if doing so allows us to keep a precious belief system intact.

This is what makes climate change the modern equivalent to Galileo's discovery that the Earth isn't the centre of the universe. For two centuries and more, fossil fuels have been at the centre of human prosperity; directly and indirectly, oil and gas and coal have enabled everything we love about civilization: abundant food, easy travel, leisure time, warm homes, education, music, medicine — all of it. Civilization runs on energy. Fossil fuels have supplied more of it than ever before. The power structures that dominate our modern world have grown directly out of fossil fuels. Now all of that is changing. A lot of people don't want it to.

It isn't just the rich and powerful (who have the most to lose) who are unhappy, either. Their misinformation has found a willing audience in the general public. After all, our challenge is far greater than Galileo's. What did it cost him to realize the Earth moved round the sun? Did it mean he was a bad person? Did he have to change the way he got to work, or what he ate, or how he lived in any way? No. But to acknowledge climate change means we also acknowledge that practically everything about the way we live must also change. It means we have to give up our gas-guzzling cars, stop flying around the world for fun, eat a lot less meat. Perhaps even more painful than changing our behaviour, acknowledging climate change means acknowledging our guilt. It means admitting all the damage we've inflicted. That damage is being done not just to nature but also to our fellow humans, because scratch the surface of fossil fuel's legacy, and you see a battlefield strewn with dead bodies. Very few of them are white.

When I start talking like this, my wife calls me Dr. Doom and reminds me of something essential: There's nothing wrong with a little cognitive dissonance. We don't have to make it all go away. We can sit with it awhile, let truths and behaviours that conflict with one another be what they are. I'm no climate monk. I avail myself of many joys and conveniences of industrial civilization.

Rather than wallow in guilt, I try to be honest, which is quite a bit more difficult.

It is, I think, the people who try to squash the dissonance entirely who come to the strangest conclusions. Instead of cohabiting with uncomfortable truth, they twist logic and reason to impossible ends, until voila! *It was antifa who started those forest fires. It was Trump who won the election. Democrats are satanic pedophiles.* When these kinds of statements are made out loud, it falls on all of us to state the obvious: They are fantasies. Believing in them is a choice. That choice inflicts harm on others. In order to avoid admitting they are playing make-believe, deniers and liars of every persuasion can always find solace in the crowd, because surrounding yourself with people who share your delusion is a reliable way of eliminating dissonance. As Festinger observed way back in 1957, "If everyone believes it, it most certainly must be true."

The maddening thing, of course, is that's just what they say about us. Everything we observe about them, they'll claim to observe about us. Don't let the false symmetry fool you. Hold as fast to the truth as you can. It's slippery but real. We need it more today than ever.

For the past five years and more, truth has seemed to be sliding from our collective grasp and falling beneath the waves. On closer inspection, it appears rumours of its death were exaggerated.

THE UNBEARABLE WHITENESS OF BEING (AN ENVIRONMENTALIST)

SOONER OR LATER, THE QUESTION ARISES: How can I justify worrying about atmospheric concentrations of invisible gases or the Pacific Ocean's pH level while so much goddamn human suffering persists? Is my capacity to help not finite? Is there not a limit to the number of societal problems to which I can dedicate time and attention? And given this, shouldn't I, like anyone with a moment to spare, be sparing it on behalf of my fellow humans who continue to endure neglect and persecution?

It's not as though I have to look far for examples. Even here, in one of the wealthiest of nations, torment is ubiquitous. I see it wander past my window every day. I live a short walk from downtown Vancouver, well within the radius of a housing crisis that afflicts thousands of people. Some of them sleep in the alley below my home office; others forage in the alley's garbage and compost bins. They are often visibly agitated, in a clear state of emotional and cognitive anguish. Every one of them has been failed by our society in a way that is literally painful to contemplate. Shouldn't I channel my altruism toward finding housing and food and access to care and whatever else is required to give them not just material sustenance but also love and dignity? Shouldn't I be writing about that instead of climate change?

You don't have to live in east Vancouver to find ready examples of injustice. It's everywhere.

Where to begin? Over 200,000 Canadians experience homelessness over the course of an average year. Even before the pandemic struck, nearly one in five Canadian children were living in poverty; that figure jumps to one in three for racialized children, 41 percent for First Nations children living off reserve, and over half (53 percent) of First Nations children living on reserve. Indigenous children also make up over half the kids living in foster care (52 percent in 2016, when the last census was taken), despite comprising 8 percent of Canada's child population. There are roughly two thousand hate crimes committed each year across the country, against Jews and Muslims and gay men and Indigenous women and just about any marginalized group you can name. Black Canadians, who make up 3 percent of the population, comprise nearly 9 percent of the prison population; for Indigenous Canadians, those proportions are 5 and 30.

Some argue that class has more influence on a person's well-being than ethnicity in this continent. A person born rich and Brown, they say, has better odds of success than a person born poor and white, as though the existence of one kind of problem (class disparity) somehow nullifies another (racism). More fundamentally, that argument ignores the fact that poverty itself is colour-coded. According to the non-profit Canada without Poverty, one in five racialized Canadian families were living in poverty as of 2020, compared to one in twenty non-racialized families.

These numbers only begin to hint at an appalling social crisis playing out right under the noses of every Canadian who isn't its victim. In every instance, the suffering is amplified by a colonial legacy of systemic racism. That legacy isn't the only driver of oppression, but it is so pervasive that no account of inequality or injustice is complete without it — especially when that account comes from a white heterosexual male, the very type of person this system is optimized to benefit.

There's just no getting around it: Genocide and slavery are the foundational sins of what we now call Canada and the United States of America, and the survivors are still dealing with the fallout. It follows that those of us who've profited from the spoils of white supremacy have a moral imperative to stamp it out.

In light of all this, worrying about things like methane leaks from distant pipelines or the pine beetle infestation ravaging the boreal can come off as frivolous. Nor is the case for environmentalism bolstered by the fact that North America's environmental movement has its own tradition of white supremacy (more on that in a moment). But scratch the surface just a little and you'll see that social justice, here and everywhere, is intimately bound up with questions of land and ecosystems, a relationship that even white environmentalists have finally come to appreciate.

It's not that human suffering shouldn't occupy the top tier of our moral attention. Nothing can be more urgent than human rights, not even climate change. It's just that framing the question as "Which problem matters more?" is the wrong way to look at it. Ecological collapse is a fundamental chapter in the story of human oppression. When two problems are intimately linked, their solutions are, too. And that's the thing about modern environmentalism: It's a powerful weapon in the battle against white supremacy.

←∿→

The idea that ecological and economic well-being are intimately connected is something even mainstream politicians now acknowledge. "The only way to build a strong economy for the future is to protect the environment at the same time" is now the official line, coming straight from the prime minister's mouth, in this case in December 2020 when he announced his plan to ramp up the tax on carbon from $30 to $170 per tonne over the next ten years. Gone (or at least going) are the days when we had to suffer the false choice between environment and economy. Since the moment he took office, Justin Trudeau has been reminding Canadians that the two "must go hand in hand."

Talk, of course, is cheap. But for this kind of talk to have risen to the level of Justinian platitude is, in and of itself, worth a moment's celebration. However obvious it may seem that natural resources are necessary for a resource-based economy, or that a people who lack clean water and air and reliable weather patterns are unlikely to flourish, it took a long time to get here. But a closer look at the details of this marriage of environmental and economic outcomes reveals an assumption that doesn't always pan out. Namely, that a country's economic health trickles down to social equality. If the last few years have taught us anything, it's how much inequality can worsen in the midst of a booming GDP.

This is a point that environmentalists and the social justice community now both embrace. They know, most of them anyway, that they're in the same fight. Both groups are demanding that the richest, most privileged members of society — white men — begin to share and make do with less: less profit, less consumption, less political and economic power. This ongoing merger of movements has its own nomenclature, with expressions like "environmental justice" and "just transition" at the top of the lexicon.

If a politician invokes the phrase "just transition," they're signalling that they understand that transitioning our economy off fossil fuels presents a threat to those hundreds of thousands of people who make a living in that industry. They're saying they take those workers' welfare seriously, and not just because they're compassionate but because they're strategically intelligent — they recognize that oil and gas workers vote, too, and can really jam the machinery of change if you don't make change appealing. Trudeau is well aware of this; in 2019, he campaigned on a promise to introduce a Just Transition Act, "ensuring that workers have access to the training and support they need to succeed in the new clean economy." Soon after he won the election, the pandemic arrived, and the act dropped off the radar.

Talk about "environmental justice" and progressive types will hear you acknowledging that the pursuit of ecological sustainability doesn't automatically reduce inequality and racism, any more than a humming economy automatically reduces poverty. Instead, you must explicitly work racial justice policies into any environmental policy, and vice versa. Again, this isn't strictly a matter of moral rectitude; it's an effective strategy for gaining broad support. When you harmonize environmental and racial policy, the likelihood of both being improved is higher than if you addressed either one in isolation.

It has to be said that it took the (overwhelmingly white) environmental community a lot longer to realize this than those who experience social injustice, for whom it has long been self-evident. For people oppressed and dispossessed, the struggle has always been one of life and death, a clarifying state of existence that white environmentalists have only recently begun to internalize.

The man known as the father of environmental justice is Robert Bullard, a Black American author and distinguished professor of urban planning and environmental policy at Texas Southern University in Houston. As Bullard puts it, "America is segregated and so is pollution." Bullard came to this issue in the late 1970s, when his wife, the lawyer Linda McKeever Bullard, asked him to help her prepare a class action lawsuit for residents of a Black middle class neighbourhood in Houston where the city had approved a landfill. When Robert Bullard dug into city records, he learned that all five of Houston's landfills were in Black neighbourhoods, despite Black people only comprising a quarter of the population.

Six of the city's eight incinerators were also in Black neighbourhoods. Eighty percent of the city's waste was dumped in Black
neighbourhoods. Seven years later, they lost the case, which, as
Bullard recalled in a 2020 interview with *Inside Climate News*, was
presided over by an openly racist white judge.

Bullard kept fighting, and writing, eighteen books so far. In
1991, he co-founded the National Black Environmental Justice
Network. All through the country, from Louisiana's "cancer alley"
to the sprawling Chevron refinery built smack in the heart of
south Los Angeles, Black and other racialized communities are dis-
proportionately forced to live beside toxic waste; they're the ones
who suffer the worst health impacts, from asthma to cancer to
premature death. The same holds true in Canada. A 2020 report by
the UN special rapporteur on human rights and toxics found that
one million low-income Canadians live within one kilometre of a
major industrial pollution source; within that demographic, racialized people outnumber non-racialized by almost two to one.

Environmental racism's most frequent targets in this country
are Indigenous. In the Ojibwe community of Grassy Narrows in
western Ontario, 90 percent of the population has symptoms
of mercury poisoning, the legacy of an upstream pulp and
paper mill. In Nova Scotia, the Mi'kmaq community of Pictou
Landing spent more than fifty years fighting to shut down a
pulp and paper mill that dumped ninety million litres of contaminated effluent per day into the tidal estuary, destroying the
aquatic ecosystem and putting an end to the local fishery. Cree
communities living downstream of the oil sands' tailing ponds
in northern Alberta have documented elevated rates of cancer,
renal failure, and lupus. In Vancouver, the Musqueam Nation,
whose reserve is at the south end of the city, has spent decades
living with the intolerable stench of a sewage pipe that off-gases
the excrement of nearly a million Vancouverites directly into
their neighbourhood.

· These are just a few examples of a ubiquitous Canadian story. But terrible as they are, the real issue goes far deeper than localized pollution. Here in Canada, it's no coincidence that the people who suffer the worst human rights abuses — whose access to healthy food and clean water and proper education and employment and shelter has been systemically withheld — are the ones who had every square inch of their land taken away. That theft is Canada's historic crime, and America's, too. It's the source of our nations' great wealth, and the cause of ongoing penury among those Indigenous nations who had their entire land base stolen. We gesture toward repentance, some of us, in the land acknowledgements that have become customary introductions at many public gatherings. But there's been no formal state apology, let alone any systemic attempt at reparations. To the contrary, every single effort made by Indigenous communities to regain the tiniest fraction of their traditional territories has been, and continues to be, regarded as an existential threat by the governments of Canada and the U.S., to be countered with the full force of the laws that they themselves have written.

It doesn't take much effort to see the connection between land and social justice in this state of affairs. If anything, it takes effort *not* to see it. And that particular effort, I feel compelled to acknowledge, is one the environmental movement has a history of making.

↞〰↠

In his 2017 essay "Canada's National Parks Are Colonial Crime Scenes," the Kwantlen journalist and author Robert Jago laid bare the terrible social cost of Canada's earliest environmental impulse.

"Canada's Parks Departments have treated Indigenous peoples like an infestation ever since the founding, in 1885, of Banff

National Park," Jago writes. Not only were the "Indians" forcibly evicted, they were also barred from returning for any reason. Jago quotes George Stewart, the park's superintendent at the time, who complained of the Indigenous presence in Banff that "their destruction of the game and depredations among the ornamental trees make their too frequent visits a matter of great concern." And so the Blackfoot Confederacy, the Tsuut'ina Nations, and the Stoney Nakoda Nations were banished from their territories, as were the Indigenous occupants of every national and provincial park that sprouted across the country in subsequent years. Not content to stop there, Canada's legislators decided the land's original inhabitants should be confined to isolated patches of unproductive earth, the more remote the better, which they could only leave with special permission from the local Indian Affairs officer. This reservation system — a sort of *Black Mirror* inversion of the national park system — became a political tourist attraction, drawing legislators from South Africa who studied and then used it as a model in creating the infrastructure of Apartheid.

You won't see any of that context in a Group of Seven painting. Nor will you see the rich Indigenous history that preceded the colonial crimes Jago describes so painfully in his essay. "The places Canada has made into parks are filled with our stories," Jago writes. "Every mountain, every valley has a name and a history for Indigenous peoples. It is in these places that our history is alive: our Mecca is here, our Magna Carta, our Thermopylae."

In the 1930s, while White Afrikaners were touring Canada, a newly minted Nazi party began studying the nineteenth-century U.S. legislation that created the legal architecture for America's pursuit of genocide. In the U.S., the project was described as "pushing the western frontier." The Germans boiled it down to "Lebensraum," and their substitution of Jews and Poles and Czechs for Sioux and Shasta and Mojave made the horror of its execution abundantly clear to most Americans. What it didn't do was cause many of those same Americans to reflect on their own historical embrace of the very same policy.

One American whom Adolf Hitler particularly admired was Madison Grant, a Manhattan aristocrat born in 1887. Grant was a founding father of environmental conservation in the United States. He belonged to a circle (along with his friend and political ally Theodore Roosevelt) that established America's constellation of national parks, forests, and game reserves. Grant also was also an ardent defender of the American bison and California redwoods. But he was better known for his 1916 book, *The Passing of the Great Race, or the Racial Basis of European History*, which Hitler described as "my bible" in a fawning letter to the author. Grant's book cast Nordic peoples as the natural leaders of humanity, pale and noble and rational, better at self-government than the darker Mediterranean peoples (the rest of humanity didn't even merit Grant's consideration). The preservation of all that was noble in the human spirit, Grant believed, depended on maintaining its whiteness, and the preservation of ecosystems likewise depended on keeping dark-skinned humans out.

If only it was just a few bad apples. But the man chosen by President Roosevelt to head both the Forest Service and the National Conservation Commission was Gifford Pinchot, a dedicated member of the American Eugenics Society, which in turn was founded by Henry Fairfield Osborn, a dedicated naturalist who also headed the board of trustees for both the American Museum of Natural History and the New York Zoological Society. For men like these, "It was an unsettlingly short step from managing forests to managing the human gene pool," as Jedediah Purdy put it in a 2015 essay for the *New Yorker*.

One of the most celebrated environmentalists in North America is John Muir, founder of the Sierra Club. Muir was a poet and a wanderer, an early model for later generations of counterculture heroism. It was Muir's literary activism, as I learned in university, that gave us Yosemite, Grand Canyon, Mount Rainier, and other dazzling American national parks. He spent his life exploring continental America on foot, developing a language of the sacred that he brought to bear in his prolific nature writing. "The clearest way to the universe is through a forest wilderness," he wrote. What I wasn't taught was how contemptuously he regarded Indigenous Peoples in his early years. Muir described the Cherokee homes he came across in the Smokey Mountains of North Carolina as "the uncouth . . . wigwams of savages," and compared them unfavourably to the settler homes of white men that were "stamped with the comforts of culture and refinement." Muir didn't know, or didn't care, that those "savages" were the descendants of a nation evicted by the U.S. army in 1832 in abrogation of their land treaty, a treaty the U.S. government ripped up the moment gold was discovered on Cherokee land. The Cherokee Nation was then forced to march over two thousand miles in the middle of winter to Oklahoma, a journey today commemorated as the Trail of Tears, because four thousand people did not survive it.

Later in his life, Muir had a change of heart. His travels brought him to Alaska, where he spent time living with the Chukchi and Tlingit Nations and finally discerned a kindred spirit among the Indigenous communities he'd earlier viewed with such disdain. But that philosophical awakening was, until very recently, the least documented in all Muir's travels, including by his own pen; his legacy remains one of general indifference to human affairs. That's more than can be said for most of the other foundational environmentalists I was taught to revere, who took a much more active role in racial oppression: John James Audubon, founder of the Audubon Society, enslaved people all his life. Aldo Leopold, legendary author of the environmentalist bible *A Sand County Almanac*, worried that immigrants would "overrun the country." The list is endless, the message clear: "Wilderness" and "nature" as we talk about them today are the artifacts of men for whom the conservation of an ecosystem was inextricably bound up with the preservation of unequal race relations. For a place to be natural and wild meant it was free of humans — especially humans of certain pigmentations.

Later generations of environmentalists — people like me — might abhor our forebears' unabashed racism. (In the wake of 2020's Black Lives Matter protests, the Sierra Club issued a public apology for its founder's racism and promised to address systemic racism within the organization, prompting other environmental groups to follow suit. "A good first baby step," said Robert Bullard, adding, "In my opinion, none of them have taken a strong stand in the way their white privilege sucks up damn near all the green dollars from foundations and donors, away from people of colour.") But we've inherited their world of ethnically cleansed parks whether we like it or not. More subtly and pervasively, we've also inherited their language, which infuses our descriptions of the environment with a vocabulary of eugenics: we elevate purity, admire supposedly pristine ecosystems and unblemished landscapes, untouched and untarnished by human hands.

↞∿↠

There's a deep irony in these white men striving to protect wild spaces by forcibly removing the very people who know how to take care of them. A century later, after all that's been lost, we're only just starting to address that irony. In 2017, the United Nations Environment Programme reported that the world's Indigenous Peoples — some 370 million people — "own, occupy or use up to 22 percent of the global land area, which is home to 80 percent of the world's biodiversity. Often overlooked by governments, their role in safeguarding territories from environmental degradation has largely gone unnoticed and undocumented until now."

In Canada, First Nations have for some years been the most powerful force in environmental protection. No other group even comes close. Non-Indigenous environmentalists and concerned citizens weigh in on major new projects, from pipelines to mega-dams; we speak as interveners at public hearings, or write clever books, or wave placards outside Parliament. But only Indigenous Peoples have the force of law on their side.

Aboriginal law, as it's still formally known, is both the newest and fastest-evolving section of Canada's legal code. If you ask me, it's also the most interesting. It captures the way we're just making this all up as we go, as well as how Indigenous perspectives are finally working their way into Canada's legal architecture.

Consider that the entire field of Aboriginal law stems from a single sentence in our Constitution, known as Section 35: "The existing aboriginal and treaty rights of the aboriginal peoples of Canada are hereby recognized and affirmed." Because the treaties were written in the vague, archaic language of the seventeenth and eighteenth centuries, and because many were arguably signed without informed consent, and because they were subsequently ignored or abused by the Canadian government (for instance, by responding to the promise of public education written into many treaties with the residential school system) — because of all these gaping uncertainties and acts of bad faith, it's terribly unclear what "the existing aboriginal and treaty rights" are exactly. That's what the courts have been spelling out, precedent by precedent, ever since Section 35 was drafted in 1982.

Section 35 does include a crucial addendum: "For greater certainty, in subsection (1) 'treaty rights' includes rights that now exist by way of land claims agreements or may be so acquired." This sentence, which reads like an afterthought to an afterthought, became the foundation of Indigenous land defence in Canada, especially in British Columbia, where almost no treaties were ever signed. (Except for a small handful of treaties on Vancouver Island, the entire province was simply taken without even the veneer of a legal contract.) As a result, Section 35 has had tectonic repercussions not just for Indigenous communities but for the environmental movement as well.

It was Section 35 that sparked the biggest environmental protest in Canadian history, the "war in the woods" to protect the old-growth forests of Clayoquot Sound on the west coast of Vancouver Island, which culminated in the arrest of more than eight hundred protesters in 1993. That drama was first sparked a decade earlier by the plan to log Meares Island in the traditional territory of the Tla-o-qui-aht First Nation, for whom the ancient cedars and spruce trees — many well over a thousand years old — made the island one of several sacred sites within Clayoquot Sound. MacMillan Bloedel (the logging company licensed by the province to clear-cut Meares Island) and the provincial government thought the Tla-o-qui-aht were being ludicrous when they hired a lawyer to argue that Meares Island was legally theirs. This was 1984, when Section 35 was two years old, and the concept of Aboriginal Title was unheard of. Up until then, industry and government logged and fished and laid pipelines wherever they wanted, without a moment's thought about anyone's consent but the Crown's. But the court ruled that Section 35 did indeed give the Tla-o-qui-aht the right to pursue Aboriginal Title to Meares Island; furthermore, until ownership of Meares was established, nobody would be allowed to log it. In response to that injunction, MacMillan Bloedel accelerated old-growth logging elsewhere in Clayoquot Sound, which led to the historic protests that finally reverted control over Clayoquot Sound to the Tla-o-qui-aht, Ahousaht, and Hesquiaht First Nations — sort of. Almost forty years later, the Tla-o-qui-aht still don't have formal title to their own land, but Meares Island remains uncut, protected by that 1984 logging injunction, and Clayoquot Sound is now a UNESCO World Heritage site.

So far, out of the 634 First Nations in Canada, only one — the community of Xeni Gwet'in, which forms part of the Tŝilhqot'in National Government in central B.C. — has regained Aboriginal Title to a portion of its traditional territories. It took them twenty-five years in court to get it. That tells you why it doesn't happen more often; these legal imbroglios have burned through the time, talent, and financial resources of entire generations of Indigenous communities. But the wave of legal victories is building, and it's already had a profound effect on the landscape of First Nations rights and environmental struggles.

Section 35 is the reason we now speak of the government's duty to consult First Nations on any industrial project that affects their traditional territory. Wherever a First Nation can demonstrate a reasonable claim to the land, a claim they may one day argue in court, federal and provincial governments are legally compelled to gain their informed consent before licensing a major industrial project. That's why the Northern Gateway pipeline project got cancelled. That's why the Trans Mountain pipeline expansion project was shut down for almost two years, forcing the government to substantially enhance its environmental safeguards. The Coastal First Nations who tried to quash the project entirely may have failed to do so in court, but they came a hell of a lot closer than any of the other groups who tried. That story isn't over, either. For now, it's moved outside the courthouse, where it still has more power than any "purely" environmental protest ever will.

The environmental movement is well aware of this. That's one of the reasons why so many of Canada's environmental campaigns centre Indigenous voices. Non-Indigenous environmentalists aren't always the best allies, and lord knows it took us too long to get here. For many years, and sometimes still today, predominantly white non-profits and journalists and politicians have exploited First Nations' unique standing (and precarious finances) in order to advance their own agendas. It's not for me to say how far non-Indigenous environmentalists have come or have yet to go, though I do think it's accurate to note that we recognize legal and moral authority when we see it.

Perhaps the best way to put it is how Kathryn Teneese, chair of the Ktunaxa Nation Council in southern B.C., once put it to me. In 2018, the Ktunaxa were fighting to prevent a massive ski resort from being built on long-sacred ground; a broad coalition of environmental groups were also opposed to the resort, though they valued the region for being prime grizzly bear habitat and an unparalleled back country ski destination, too. "We may be walking in the same direction," Teneese said when I asked about her relationship with those non-Indigenous environmentalists, "but we're not holding hands."

↞∿↠

It's common to hear the word "apocalypse" in environmental conversations. It probably gets used too often and too lightly, without enough thought for those peoples among us who have survived a more literal brush with apocalypse than many of us dare to imagine.

We ignore those survivors, and the justice they're owed, at our peril. One of the lessons we can learn is that a people's relationship with the land is no abstract consideration but a matter of life and death. Slavery and genocide, two distinct offspring of North America's colonial history, have at least that much in common. Both are predicated on the violent uprooting of whole cultures from their ecosystems.

Acknowledging this reality isn't just about atoning for the past. It's about committing to a shared future in which land and leadership are equitably distributed and human well-being is explicitly tied to environmental justice. The Anthropocene has begun, a sixth great extinction is underway. How many species will it claim? Will our own survive? The answers may depend on a different question altogether, more practical than is often meant: How can we live with ourselves?

LET'S GET DRUNK AND CELEBRATE THE FUTURE

I HAVE BEFORE ME A number of hardcover tomes, each one thick enough to stop a bullet, and I was wondering if you would like me to boil them down to a lengthy data-driven explanation of why pursuing infinite economic growth on a finite planet is a nightmare dressed up as a dream, or should we just get drunk instead and celebrate the future?

Humans have been consuming alcohol for at least ten thousand years. If that isn't sustainable, I don't know what is. Why don't any of these manifestos mention this? Our ancestors drank and so will our descendants. The chain of inebriation connecting our past to our future is but one of many fine portents worth toasting, and so, in the spirit of kicking things off, here's to the eternal growth of grapes and hops and juniper and agave, and to the alchemy of fermentation that transforms our souls but leaves the Earth alone.

You know what goes well with a drink? Music. Which also imposes no toll on sky or sea or forest, and also connects our past to our future, and is impossible to overconsume. Here's to melodies and the people who make them.

Combine alcohol and music and you get another ecofriendly phenomenon: dancing. The biosphere will support these three wonderful things in perpetuity. We never have to give them up, or make do with less of them, we can just always have more and they can grow and evolve and get better and better and reach more and more people and so let's raise our third glass — the best of the evening — to this holy trinity.

Remember, despite all there is to mourn, we are celebrating. Let this bacchanal be positive. No angry drunks. Licentiousness is encouraged, within the boundaries of consent, of course. For here is yet another cause to celebrate: the rise of prophylactics, which, you must admit, go well with alcohol and music and dancing. Much better than the terrible tomes on my desk, all of them urging me to wag my finger and declare that it's time to get serious, face reality, learn to do more with less. Time to lament the end of a golden age of economic growth and glorious overconsumption because the world has reached a hundred tipping points all at once and there just isn't —

No! We can't subvert the laws of physics, but we can transcend one or two of biology. We, alone among species, have smashed the link between sex and reproduction. We can fornicate for fun, no offspring attached. Of course, unwanted pregnancies do still occur, but already so much less often than before, and anyway — remember? — we're celebrating the future, when everyone's rights to reproductive freedom and bodily autonomy are protected. Specifically, we're celebrating an unprecedented occurrence that is just around the corner: For the first time in the history of life on Earth, our species is about to consciously and voluntarily *stop multiplying*. Toast number four is to flipping Malthus's model on its head: Exponential population growth was itself the limit, and we have overcome it without resorting to prudishness.

Of course, you can't celebrate the future without acknowledging the past. Over the course of our debauchery, some nostalgia may arise. It's better than forgetting. We can play slow songs and hold each other close as we remember the animals and plants of the past and, sadder still, those soon to be past. It's only right to mourn them and the holes they leave behind in the landscapes they once graced. We'll need some Irish drinking songs, may well need every Irish person to bring their own tune, one for each species that will, for the rest of our existence, stalk only the pages of storybooks. The right whale and the southern mountain caribou, the harlequin frog and the tiger beetle, the western prairie fringed orchid that we'll never get to give our lover on her birthday. I pour one out for each of you. I see you through my tears. I promise to remember you. We never met and never will, and I don't know what's sadder.

But these authors on my desk who talk about the End of Growth, the Rise and Fall of American Growth, How Growth Became the Enemy of Prosperity, or simply spend five hundred pages charting the logarithmic curves of everything from micro-organisms to megacities — these guys are neither mourning nor celebrating. They're just sort of hunkering in, the way you would for winter in Siberia. Good luck sparking a movement, fellas. Excellent writers and researchers all, don't get me wrong, collec-tively providing thousands of pages of necessary historical fact and unimpeachable argument to prove truths so obvious they've become invisible, like the lunacy of interpreting Moore's law to mean we'll one day live forever, which is something highly intelli-gent people sincerely believe. We won't! We're all gonna die, even the first trillionaire.

What these buzzkills need is panache, something the pur-veyors of growth and infinity possess in abundance. Let's take it back, and sprinkle in some imagination, another unlimited resource. Like how about the death of the *last* trillionaire. Can we celebrate that?

Say it together: One day, only children will dream of becoming insanely rich. One day, instead of trying to mound up mountains

of treasure like evil dragons and Disney villains, adults everywhere will conspire to amass friends and meaningful memories.
They will compete to memorize more poems than anyone else
in their neighbourhood, tell the funniest jokes in the city, remember
their literal dreams in the night. They will expend vast energies on
plumbing the strange miracle of life becoming aware of itself, also
known as human consciousness.

Let's get drunk and celebrate that future. It's only one of many
on the menu, so it's important we drink to its success; otherwise
it may vanish for lack of demand. Yes, yes, it's true, that dream
we had of always having more of everything — poof. It's gone,
in material terms. We can't have four cars each, even if they're
all electric. There's not going to be enough room for ten billion
of us to each have our own five-bedroom house with a yard and
a barbecue; we won't all be able to eat two cows this year and
three the next. But not everyone has to stop having more. Not by
a long shot. Those millions and billions still stuck on the other side
of having enough, *they* can keep growing. For them, for years to
come, the mantra remains unchanged: more jobs, more houses,
more food, more schools, more clothes.

But us? The well supplied but never sated? We already have
what they're after. Over here, economic growth is a diversionary
tactic. The forty-four richest people in my country made $64 bil-
lion last year. We have some distribution issues to take care of, but
one thing we're not lacking is enough.

Here's to realizing that, and the liberation it will bring. All that
creative energy and intelligence and ambition — free at last! Put it
wherever you want, other than money and *things*! Music, mountain climbing, gymnastics, kung fu, dinner parties, mathematics,
physics — in the future we'll still need most of the professions
we have today, the chemists and the engineers and architects, and
the ones we lose will be replaced by others we can't even picture
yet. The ones we'll need most are economists, for they will be

charged with answering the question: How do you run the world on enough instead of more?

That's the question they always ask when I raise this subject, anyway. When I say, *I don't know!* they take my non-answer as proof that the question itself is absurd. And so I raise a sixth glass to your logic, economist friends, and quietly note that even the laborious textbooks staring gloomily at me never go far beyond diagnosis. What a challenge! The global economy was built on growth, growth is in our DNA, we're all really going to have to lean into this one. But remember, we love challenges! That's also in our DNA. This is just a new one. A seventh toast to wash the bitter taste of indignation from my mouth after that last one: to the sparkling unfamiliarity of circular economics!

Now let's get back to celebrating a future in which it seems hilarious that we almost ruined the climate. *Can you believe*, we'll say, like remembering that time the car flipped off the highway and everyone walked away unscathed, *how close that was?* Because — and here's another tricky bit where things could get depressing — beating climate change is going to be the easy part. That's already underway. We're trading engines for batteries, plugging our electric grids into wind turbines and solar farms instead of coal piles. It's happening, but it's such a big project that it's blocking our view of the bigger one right behind it. We can't yet see that huge mountain just around the corner. But we will. The moment we transcend carbon, we'll realize we're still eating up the planet at an awful, awful rate. We'll realize that even if the atmosphere's safe, the oceans and forests and wetlands and prairies and rivers and lakes and aquifers are decidedly not.

This one calls for something stronger. We may have to go beyond the usual brewmasters and vintners, reach into traditions like those of the Mazatec in Oaxaca. Yes. Saddle up for toast number eight. It's time to sip some psilocybin tea, because this one really requires a full flipping of the script. We have to eliminate some core cravings that have been with us for so long we think of them — when we think of them at all — as virtues: greed disguised as ambition, selfishness disguised as freedom, inequality described as opportunity.

But don't worry — as strange as it all looks when you step through the looking glass, it's neither strange nor new in actual fact. It's just what the poets and prophets have been saying all along. Break on through to the other side. That's where the real treasure lies.

As philosophers and musicians have always known, this needs to be said and sung and recited in as many ways as we have words to say it. Although I do fear words alone may never be enough — if these mighty manifestos arrayed before me suggest anything, it's that. You know what we need to break this spell? We need models. I mean the human kind. Come all ye beautiful people, ye minstrels and ye bards, ye ninjas and jedis of social grace, ye havers of good times and setters of trends, ye charming gorgeous influencers. Enchant us! Lure us! Tempt us into the beautiful future that awaits. Set the table, throw the party, show us where and how the good times will be had in the future — not aboard the private jet with the gold-plated interior and $25,000 bottles of Dom, nein! Aboard your homemade sailboat, maybe, with a pint of locally brewed stout in hand, ja. Atop the mountain you hiked with a friend; at your vegetarian dinner party, surrounded by friends at the full moon festival. Do whatever you have to do, just make it clear this is where the party's at, and there's room for everyone. Drive home the point that all our innermost desires are beckoning from this new future we're celebrating drunkenly: sex, love, companionship, food, intoxication, clarity, adventure, creativity, stories, and music and family and purpose and transcendence.

None of us is getting out of here alive. We agreed on that already, right? And that's okay. It's more than okay; it's exactly what gives this party its pizzazz. The mushrooms are kicking in now, and you're realizing you don't have to be so scared of death and dying. You're realizing that this fevered pursuit of growth and infinity was a weird meta-collective-self-hypnosis thing that long ago outgrew (apologies) its usefulness, became a malignant force of habit, a veil to hide our increasingly debilitating fear of death, our outright denial of death. Snap out of it. You're going to die. When the inevitable moment arrives, if you haven't joined our celebration of the future, you'll realize it passed you by. Your final moments will not be marked by the satisfaction of a life well lived or the spiritual glow of imminent revelation; no, instead you'll be consumed by the monumental folly — the horrendously squandered opportunity — that it was to spend your life pursuing *more*.

I'd like to say these things to the world's economists and presidents and CEOs, but they're smarter than me, and they're busy. Busy making money for themselves and others, busy making sure that whatever else happens, this little civilization of ours keeps growing. It's not their fault. All they see, when they have time to glance out the window, is a bunch of half-dressed idiots dancing in the street and making out. I admit it doesn't look professional. It's not going to change their mind, any more than these books in front of me did. So let's get drunk instead and celebrate the future.

> A poem knows no compromise, but men live by compromise.
> The individual who can stand up under this contradiction
> and act is a fool and will change the world.
>
> — Günter Grass, "On Writers as Court
> Jesters and on Non-Existent Courts"

REBEL, REBEL

October 7, 2019

I AM STANDING FORTY METRES above the black Pacific, one hour before midnight. The Burrard Street Bridge is empty as a movie set all the way up to the centre of the span, where two lines of people have squared off. On one side are twenty poker-faced police officers with fluorescent vests over their uniforms; arrayed against them is a bedraggled Gore-Tex rainbow of people with their arms draped around each other's shoulders, some solemn and some smiling, all singing as they sway. Half a moon has just emerged from the clouds that doused the city all day.

This art deco bridge, a main conduit to downtown Vancouver, is one of seven across Canada and several dozen more around the world that were blocked this morning by members of Extinction Rebellion (XR), the radical environmental movement that introduced itself to the world one year ago in London when six thousand activists blocked the five main bridges over the River Thames for several hours. Then as now, the disruption was meant to remind the world that the daily business of modern civilization has triggered the sixth great extinction. The Thames blockade marked the climax of weeks of traffic-jamming street theatre and civil disobedience that occupied a good portion of the international news cycle and led directly to the U.K. becoming the first country in the world to officially declare a climate emergency. Six months later, XR-UK came back even stronger, parking a pink sailboat named *Tell the Truth* in the middle of Oxford Circus, one of several key intersections throughout London transformed that spring of 2019 into a two-week Burning Man for the biosphere.

Fearless and creative and bursting with imperial panache, Extinction Rebellion's founders set a new tone for ecological concern that inspired sister chapters to sprout around the colonies at a rate of one per week. Like Occupy, Black Lives Matter, and Idle No More, Extinction Rebellion declared itself a decentralized organization with no official leaders. Anyone can start a franchise, emblazon the group's stylized hourglass on a banner or a hundred T-shirts, rouse the rabble for a march. Any day of any week is fine, but twice a year — each spring and fall — the global XR family coordinates a day of international action.

In October 2019, a period that already seems quaint, most of the bridges blocked by XR chapters around the world were cleared by the police within hours. Not in Vancouver. Here, in the city where Greenpeace and David Suzuki were born, courteous officers did the protesters' work for them, blocking off both ends of the Burrard Bridge at nine in the morning and rerouting traffic before angry encounters could erupt between commuters and rebels. The organizers had mixed feelings about this police collaboration — "Arrests are what get us on the news," one told me — though as the day wore on and word of fist fights came in from around Canada (Alberta's motorists took particular joy in the rare opportunity to punch environmentalists in the face), a consensus emerged that maybe a little municipal goodwill wasn't the worst foot to lead with.

But finally that will ran out. At ten in the evening, after a wet day of hanging banners from the bridge and singing peace songs and speaking to reporters and listening to speeches from university students and environmental activists and Indigenous elders — after twelve hours of this, and now that the reporters were all gone, and most of the protesters, too, and everything was really pretty sleepy, the order came down to clear the rebels out. Go home or be arrested. Most people obeyed, packing their signs and tents off the vehicle lanes and onto the sidewalk. But ten stalwarts decided to stand their ground. They were given four polite warnings from the senior officer, a patriarchal figure with a full head of silver hair; then the arrests began: One by one, calmly, ceremonially, having agreed from the outset not to struggle because the movement is grounded in nonviolence, the rebels either walked or (in the case of one trembling elder) let themselves be gently carried into the back of a police van. The rest of us, watching safely and legally from the bike lane, cheered the martyrs on.

All this was expected. At the information sessions held throughout the city over the previous weeks, lawyers and veteran activists advised prospective arrestees on what kind of treatment to expect from the police (cordial, unless you put up a fight), what they'd be charged with (civil misdemeanour), how those charges would affect their future travel and job prospects (lightly but adversely).

But nobody anticipated this final act. Just before the last arrest, a group of twenty rebels jumps back into the middle of the road and starts to sing. This was not part of the plan. Nobody discussed this. Slowly but spontaneously, two huddles — protester and police — spool out to form two lines transecting all four vehicle lanes. And there they pause, facing one another like partners at a line dance. One side sings, the other stares. The air between them crackles with the weird energy of unspoken surprise.

The word "weird" has its roots in the Old English word "wyrd": fate, fortune, destiny. This is either how it ends, or how it begins.

〜〜〉

I don't like protests. The chanting, the slogans, the righteous certainty; for me, these conjure every political movement in history that followed a good cause down the path to tyranny. This is a terrible attitude and I wish it wasn't mine. I'd like to kick Lenin and Castro and Gaddafi out of my head and replace them with King and Gandhi and Greta. Although Gandhi was also a problematic figure if you look at —

I apologize in advance for my faulty circuitry.

It's not that I don't believe in good causes, or join their marches when they come around. The climate strikes and Black Lives Matter marches of recent years have been a master class in people power, and I've found that immersing myself in crowds of that size helps to wash the cynicism from my soul. But anyone can join a million-person march; at that point the absurd thing is staying home. What really takes heart and a Don Quixote threshold for futile endeavour is to join a protest that's only gathered a few dozen people to shout in the wind, which, let's acknowledge, is how every movement starts and where most of them end.

It's not clear to me whether I see in activism something that I wish I had in me, or something I already have and would like to expunge. Probably both. Activism, for me, is like that family member you love but quickly get annoyed by. The passion, the willingness to put your body in service of a greater good, the disregard for the judgment of society — put these on the "wish I had it" side. Counterbalancing that is my German background, which compels me to recoil from collectively raised hands and a hollering crowd. Where some see conviction and clarity, I see complex issues reduced to good and evil. *Down with capitalism! Defund the police! No more pipelines!* These are great debate points, but not all the people making them are interested in debate. Things don't go well if I suggest that maybe capitalism just needs a tighter leash, or note there's some ambiguity about who means what by "defund," or speculate that building one last pipeline might not be the literal end of the world. True, a person can drown in nuance, and some things *are* black and white. It's just that most are not. The more I suppress such misgivings in the midst of any given protest, the more my body betrays me. My eyebrows rise involuntarily, my fist wavers in the sky. These tics rightly draw suspicious side eyes from my fellow marchers. It doesn't help that I started shaving my head four years ago thanks to a receding hairline, which, combined with my imperious nose and square Teutonic cranium, can in certain lights lend me the mien of a neo-Nazi or an undercover cop.

Jonathan Matthew Smucker, the author of *Hegemony How-To: A Road Map for Radicals*, a widely read handbook for social movements, has written about "the political identity paradox" that hamstrings many protest movements: Activist communities depend on the power of identity to form a supportive community — that's what keeps them going against the odds — but this same mutual support network can take on an exclusive air, isolating the movement from the broader public. Without public buy-in, no protest movement can make the change it seeks.

Extinction Rebellion is not immune to this, but it does have some advantages. Chief among them is an increasingly widespread public awareness that our planet is, indeed, in dire fucking straits. A great many people in a great many countries are horrified by the ecological calamity unfolding all around us. Those same people are frustrated by the lack of political will to address this crisis. An international poll released a few weeks before the Burrard Bridge blockade found climate change was the number one concern among voters in seven of the eight countries surveyed — Canada, the U.K., Brazil, Germany, France, Poland, and Italy; in the United States, climate change came in third, behind terrorism and health care.

That said, the Burrard Bridge blockade aggravated thousands of motorists who had no way of making the connection between their 'unexpected detour at rush hour and the death of pollinating bees, unless they heard about it later on the news. Even then, the dots formed an abstract picture at best. And when "disrupting business as usual" makes parents half an hour late picking up their kids from school, you start running into Smucker's warning about antagonizing the broader public.

Is it possible, or even desirable, for Extinction Rebellion to resolve that fundamental tension? The whole point of civil disobedience is to poke a stick in the status quo. If you're not pissing someone off, you're not doing it right. The question then becomes *who* are you trying to goad. "We're just a ragtag bunch of nobodies up against the federal government," one of XR-Vancouver's organizers, a thoughtful woman who was finishing her PhD in sustainable agriculture, told me many months later, when we were seated in a forest underneath a protester suspended twenty metres up in the canopy of hundred-year-old cottonwoods. She added, "It's insane."

An impressively honest assessment. But is it government we're up against, or is it our own human nature, expressed in Canada by one of the most representative governments on Earth? She believed it was indeed the government, that our leadership has been co-opted at every level by short-sighted financial interests that thwart at every turn the will of a people yearning for sustainability. I honestly couldn't tell who was more naive — her for believing the masses truly want an ecologically attuned way of life, or me for believing our government isn't totally captured by the extractive sector. She was intelligent and committed, that rare person who acts on what she thinks and feels with enough authority to make doubts like mine sound a lot like excuses.

I spent a year tagging along with Extinction Rebellion's Vancouver group. "Lurking at the fringes" is probably a better way to put it. I introduced myself to everyone I met as an environmental writer who was interested in writing about this movement.

They welcomed me with a wariness that added to my respect. The relationship never got past the awkward phase. I attended some meetings and street-theatre actions, lent my voice to their chorus, but refrained from anything that could be called active involvement. I wouldn't even hold a sign on the (true enough) grounds that I wanted to be free to mill about, talk with other protesters and passersby. Nobody challenged me; everyone was polite. At the end of the year, I learned many in the group assumed I was an undercover cop.

Why did I want to write about Extinction Rebellion? Was it only to criticize or mock them? If so, better to ignore them. But no. There was something about them I admired, along with something I distrusted, and the tension between those two qualities drew me in. Many well-respected public figures have cheered for Extinction Rebellion. People like George Monbiot, who speaks at their rallies and gets arrested with them, and Greta Thunberg, and Christiana Figueres, former head of the United Nations Framework Convention on Climate Change and a key architect of the Paris Agreement, who has written of Extinction Rebellion that "civil disobedience is not only a moral choice, it is also the most powerful way of shaping world politics."

Extinction Rebellion embodies the heroic attempt in all its earnest romanticism. They risk jail and humiliation for the sake of a dying planet. It's more than I've ever done.

I can imagine a twenty-year-old version of myself discovering Extinction Rebellion and saying to himself, *At last, someone dares to dream.* They're not trying to stop a pipeline or a coal mine or a fish farm. They're not aiming to get the Forest Practices Code rewritten. They're trying to inject society with the urgency of the moment.

And there's the heart of the conundrum. It applies not just to members of Extinction Rebellion but to anyone on Earth who contemplates the madness of a civilization hurtling toward its own undoing. You're crazy to try and stop it, and crazy if you don't.

↞∿↠

Extinction Rebellion has three formal demands:

1) Tell the Truth: Government must tell the truth about the climate and ecological emergency, working with other institutions to communicate the urgency for change.
2) Act Now: Government must act now to halt biodiversity loss and reduce greenhouse gas emissions to net zero by 2025.
3) Beyond Politics: Government must create and be led by the decisions of a Citizens' Assembly on climate and ecological justice.

Since these demands were put to paper in 2018, many governments around the world have followed the U.K. in declaring a "climate emergency," Canada included. Whether or not Extinction Rebellion had anything to do with these official declarations, at least they can claim some collateral success.

But nobody's declared an "ecological emergency," and that's where this list of demands goes beyond wild optimism and enters a realm where the laws of physics no longer apply. Halt biodiversity loss by 2025? There are *one million* species on the brink of extinction — our industrial momentum is such that the only way to prevent many of those species from disappearing over the coming decades would be to move humanity to Mars next year. Get to net zero greenhouse gas emissions also by 2025? In early 2020, we shut down most of the global economy and emissions dropped by less than a fifth. For a month or two. Install a citizens' assembly to oversee these impossible timelines? We already have a citizens' assembly, one that cost millions of lives and took several centuries to install. It's called democracy.

If you're feeling charitable, you could interpret Extinction Rebellion's demands as an attempt to move the Overton window — the range of ideas that are considered acceptable in public discourse, ideas that a politician could endorse without getting kicked out of office. Joseph Overton, the late policy thinker after whom the term is named, was a libertarian who saw extreme positions as an effective lever for moving the needle of public opinion. In sales terms, you start by asking twenty thousand for the used car you're selling; then when you cut the price in half, it sounds reasonable, even though it's only worth five grand.

In this way, Extinction Rebellion's demand to bring carbon emissions down to zero by 2025 makes cutting emissions in half by 2030 (the goal set by the Intergovernmental Panel on Climate Change) sound pretty doable.

But Extinction Rebellion isn't trying to move the needle. They hate needles. They reject incremental progress, even when it comes in big increments. The rebels I spoke with earnestly believed in the literal necessity of their demands. Their theory of change didn't come from Joseph Overton but from a political scientist at Harvard University named Erica Chenoweth.

It was Chenoweth who came up with the 3.5 percent rule for successful rebellion. In 2006, Chenoweth was completing their PhD in political science at the University of Colorado when they attended a four-day workshop hosted by the International Center on Nonviolent Conflict. Chenoweth was highly skeptical of the workshop's message that nonviolent resistance is the best way to overthrow a government. "I took for granted, as did all the political scientists I was familiar with, that the serious thing, the thing you do if you're a rebel group that really wants results, is you take up arms," Chenoweth recalled for a *New Yorker* profile published in 2020. But neither Chenoweth nor the workshop leaders had any numbers to back up their assumptions; it turned out nobody had ever done a systematic analysis comparing the outcomes of violent and nonviolent uprisings.

After that workshop, Chenoweth decided to run the numbers. Together with Maria Stephan, a director at the International Center on Nonviolent Conflict who also attended the workshop, they built a database of every attempted revolution on Earth between 1900 and 2006 — 323 in all, from Uruguay's Blancos Rebellion and the Greek resistance of Nazi occupation to India's independence movement and the Philippines' People Power Revolution. In 2011, Chenoweth and Stephan published their findings in *Why Civil Resistance Works: The Strategic Logic of Nonviolent Conflict*, which landed like a peaceful nuke in both academic and activist society around the world. It was the world's first statistical proof that nonviolent resistance is the most effective way to replace a government, with just under twice the success rate of armed uprisings.

Why Civil Resistance Works offers one detailed narrative case study after another, examining why some movements succeeded and others failed. There's a lot to chew over, but the thing that really grabbed activists' attention, including the founders of Extinction Rebellion who quote it all the time, was a peculiar statistic that went viral after Chenoweth revealed it in a 2013 TED Talk: the 3.5 percent rule, which found that in every case where a nonviolent uprising gathered the "sustained participation" of 3.5 percent of the population, it achieved its goal. Once that proportion of the population is in the streets, governments must respond to their demands or be replaced. (As the *New Yorker* profile pointed out, since 2013 Chenoweth has learned of two campaigns, in Brunei and Bahrain, where this rule failed to hold true. What's more, it appears that over the last decade unpopular regimes have gotten better at quelling popular uprisings of all persuasions, a fact Chenoweth blames on the internet. But the success ratio of nonviolence over violence hasn't changed.)

The 3.5 percent rule is a cornerstone of Extinction Rebellion's strategic vision. It saturates their culture, from the founding organizers in London all the way down to casual members of branch plants in the colonies. Getting 3.5 percent of the population to join them in the streets is probably XR's most concrete goal. But what would happen if they achieved it? Let's say they did in Canada. Just over 1.5 million people join them in the streets and stay there for weeks. A problem would instantly arise: What exactly are they asking for? How do you translate XR's three core demands into policy? Creating a citizens' assembly might be doable (let's ignore the prickly details about who is appointed to the assembly, by whom, and how much power they'll wield over which aspects of our daily lives), but what about the other two? How do you legislate honesty and a halt to biodiversity loss and the complete decarbonization of the world's fourth-largest oil and gas producing country by 2025? The movements Chenoweth identified as successful had infinitely more concrete and achievable aims — toppling a government, ousting a foreign occupier, or territorial secession. It's one thing to overthrow Slobodan Milošević. It's quite another to rewrite the blueprints of an industrial economy that is inextricably bound up with the industrial economies of every other country on Earth. So far as I know, none of Chenoweth's case studies succeeded in overthrowing human voraciousness.

The people I can imagine nodding along with the above critique include the pragmatic guardians of the status quo. These are the legislators and CEOs (or their supporters and customers) who balance genuine environmental concerns against a lifetime of proof that economic growth and resource consumption are essential to civilization's well-being. They subject every prospective policy to a cost-benefit analysis and know very well how damaging radical change can be to society. Recognizing that things can't quite go on as they are, this group preaches incremental change. Its leaders are sharp and industrious and highly motivated. They rise early, shower and shave, wear clothes that fit them well and emit no bodily odours in public. You won't find bits of leaf in their hair. With the notable exception of Justin Trudeau, they don't say "um" when they speak.

They are, in other words, far more organized than the people organizing Extinction Rebellion. They have a much better grasp of how the complex machinery of society fits together. And they're leading us to ruin.

They're leading us to ruin because they have focused their efforts on maintaining the most successful economic system in the history of our species. That system operates within a time horizon measured in quarter-year units and regards the workings of the biosphere as external to the mechanisms of finance. Only as an afterthought do they train their intellects on the preservation of a global ecosystem, for the simple reason that the global ecosystem has always run itself. Unlike the global economy, no one's ever seen it collapse.

↞∿↠

November 26, 2019

XR-Vancouver's General Assembly, the first since the Burrard Bridge blockade, takes place in a dilapidated church. It's a mostly white crowd, maybe a hundred people, with an even age split from seniors to millennials and a better dress code than I expected — as many button-down shirts (albeit mostly plaid) as hoodies. Some people are new, some have been involved since XR-Vancouver first started meeting half a year ago.

The meeting opens with a land acknowledgement, delivered in heartfelt but unsentimental tones by a wiry middle-aged woman with short brown hair and a sharp nose who looks like she could catch and kill a fox with her bare hands. "We are gathered on the Coast Salish territories of the Musqueam, Squamish, and Tsleil-Waututh Nations," she says, with the professional speaker's gift for seeming to speak to each of us personally. "This isn't something we say lightly or just to get out of the way so we can move on with the business at hand. It's a profound statement of fact. We're on stolen land. There's no easy resolution here, but everything we do as an organization and as human beings has to take that into account." She reads out the evening's agenda next, then hands the mic to a solemn, dishevelled blond man who reads out a passage from *The Uninhabitable Earth*, a depressing book that spells out a series of nightmare scenarios for climate change in taxonomic detail. When he finishes the paragraph, he asks us to introduce ourselves to someone sitting near us. I say hello to the person sitting next to me, a kind-faced woman in her early fifties who turns out to be one of XR's organizers for Indigenous outreach.

I tell her I'm an environmental journalist. She says she's a linguist. We agree on the importance of word choice. I say, a bit sheepishly, that I've never been entirely comfortable with activism. A look on her face makes me want to explain. Take the people protesting the Trans Mountain pipeline project, I say. I admire them, I'm on their side, I've even joined those protests a few times. But I *can* see why Prime Minister Trudeau feels he has to support the project. I mean, how else was he going to get Alberta's support for his national climate plan, and how was that plan ever going to succeed without Alberta's support? The optics sure suck, and there's no getting around the opposition of Coastal First Nations — but in pure climate terms, I'm not sure one more pipeline is quite as big of a deal as some make it out to be. After all Trudeau *is* phasing out coal power in this country, really quite aggressively, and he's burned a ton of political capital by pushing the national carbon tax. It's not that I'm *for* the pipeline, don't get me wrong, I just don't envy the line Trudeau has to walk if he wants to take action on climate and get re-elected in an oil-smitten country like Canada.

My new friend, the linguist, takes this in with a look I'm sure I'll have on my face the first time my daughter comes home tipsy and thinks I can't tell. She says she prefers "land protector" to "protester." I agree that's a better description. She says she feels that climate change is a symptom, not the problem, that the real problem is inside of us, and here again I nod quite emphatically. She says there's a Cree word for this sickness we white people carry inside us. There isn't an Indigenous person in the whole room, as far as I can tell, but it feels unwise to say so. I figure First Nations people have their hands full enough as it is, why should we expect them to join a group like this? Would that help them, or would that simply make us feel better about ourselves?

The linguist interrupts this thought to ask how it made me feel to hear that passage from *The Uninhabitable Earth.* Terrible, I say in all honesty. I tell her I've read the book and was very moved by it. She is gazing upon me with unstated wonderment that I could read such a book and not wish obliteration upon every last pipeline and the politicians who support them.

The meeting resumes. Thank God. We in the audience turn our attention to a series of speakers. Extinction Rebellion needs to decide what the next big action will be six months hence, but between now and then, there are any number of smaller street actions to organize, including a show of support for the ten rebels who got arrested on the bridge — they have a court appearance coming up, and XR will be gathering outside the provincial court to demonstrate.

Later, when the assembly breaks up into smaller group discussions, I meet another kind and intelligent woman. She ran for provincial office in the 2013 election as the New Democratic Party's candidate for a riding just outside Vancouver. Lost by a hair. Now she's joining the other side — changing the system from without. She laughs about having tried to start an XR chapter in the suburb where she lives, but it got taken over by loose cannons. We agree: crazy if you do, crazy if you don't.

There's another person, whom I'd met on the bridge last month, a slight woman in her mid-twenties with fine brown hair tied back in a ponytail and skin so translucent it makes her intelligent brown eyes seem to glow. Her name is Lexa. We shake hands, hers trembling slightly. She looks like a mage or a prophet. She is shy and quiet to the point of invisibility, not exactly the poster child of civil disobedience. But that's not her role, she tells me. She shows up at the actions, but her real contribution is behind the scenes: Lexa does much of the research and writing that goes into XR-Vancouver's public engagement — press releases, letters to politicians, correspondence with the mothership, XR-UK, whose political philosophy drives the strategic decisions of chapters like this one.

Lexa has a degree in linguistics from UBC, and we chat about our mutual love of language and storytelling. She's been with the Vancouver group since its inception and treats it as a full-time job. She is simultaneously unlike any other rebel I meet and one of the most emblematic. She's new to environmental concerns, having grown up in a fundamentalist Christian household — "pretty much as fundamental as you get" — in which such things were never discussed. But XR's message resonates with the religious world view she grew up with. Climate change is a secular, rational apocalypse that humans could actually do something about.

"For a lot of my life I had this ideal that I wanted to go out and save the world," she says. "I had this intuition that there was something very broken about the way things were. I couldn't put my finger on it. I would have called it some kind of supernatural evil. But I had that intuition: Why are people so selfish and greedy, why does everything seem so broken?" When she broke with her faith during her university years, she "stopped thinking in terms of super-natural evil and started thinking in terms of the evil that humans can inflict on each other, and the evil that systems we create can inflict on us, even though we might not intend them for harm."

The loss of her religion put enormous strain on Lexa's relation-ship with her parents, with whom she still lived, and plunged her into a deep depression from which she still hadn't fully emerged four years later. But joining Extinction Rebellion had marked a turning point, providing both community and a sense of purpose she'd never had before. "I'm not that great at socializing or meet-ing people," she says. "XR made that a little easier. I don't know that I had experienced a good model of community before."

Lexa's university years had also introduced her to socialism and anarchist theory, ideologies that would have been anathema to the patriarchal world view she was raised with but whose egali-tarianism she now found compelling, even liberating. "I think that understanding was really what made XR appeal to me," she says, "because it wasn't just about environmental concern. It was also about this idea that had already been percolating in my mind that our democracy itself was broken, and XR was offering a solution to that — something concrete I could fight for that would bring the world closer to that political ideal that I had. And just in the way that it organized, it was very close to the anarchist ideals that I held: people self-organizing in a horizontal way and getting rid of hierarchies. All of that made XR a very good fit. It got me out of my depression; it made me feel less helpless about where I was. I wasn't just endlessly criticizing and lamenting the way the world was, I was able to do something about it."

There's one other person I meet that night: Harold, a retired Anglican minister in his early seventies. Harold is in charge of outreach, which is to say recruitment. He's the one who gives The Talk. I'd seen him, too, on the bridge last month, though we hadn't spoken then. He's hard to miss, a tall man with a shock of white hair and an open, friendly face. He exudes a boyish earnestness that has just enough cheeky wisdom in it to not seem naive. After the mingling, he leads a couple dozen of us newcomers up the stairs to the church's prayer room, where we gather in the pews and receive The Talk. There is nothing religious about it, but a lifetime of affable preaching has gone into his delivery, and I find my skepticism overwhelmed. He describes the urgency of the moment in clear, relatable terms, searching as he goes for the answers he knows don't exist, refusing to allow their absence to obstruct the quest for a proper response. The thing that impressed him most when he first encountered Extinction Rebellion, he tells us, is that everyone here knows failure is the most likely outcome, and still they press on.

↞↠

I meet Harold for coffee a short while later.

He tells me about his Anglican upbringing and the thesis in religious studies he carried out as a young man at the University of Alberta, a few blocks from where I would have been a toddler at the time. He asked atheist scientists about their views on "the big questions," to see how they thought about the miracle of existence without the guideposts of religious faith. As he'd expected, the men he spoke to were all very articulate and clear and didn't lose any sleep over questions of eternity and soul; they were focused on the hows of the universe rather than the whys. What surprised Harold were the scientists who did believe in God. They each revealed their religious doubts in certain pregnant pauses that came up when he grilled them hard enough. They were smart enough to talk their way out of it, but Harold could see they were conflicted. "I could see it, because I had my doubts, too," he says.

For Harold, Christianity lost its credibility a long time ago, when he learned about the countless abuses of power the Church has perpetrated over the years. Institutional sins aside, he'd never really believed in God as an omnipotent white guy in the sky. But that was hardly the only way to interpret the divine, and when Harold beheld reality, he perceived it as infused with divinity. What to do with that intimation? His father had been an Anglican minister, and there was enough room for doctrinal interpretation in that faith for Harold to follow in the family tradition.

Harold's feeling toward religion mirrored mine about Extinction Rebellion. They were acting on a fundamental truth — very compelling. But intellectually there were gaps.

One of the things that kept Harold from abandoning his faith was the radical social-activism of Jesus Christ. Not the Son of God Jesus, but Jesus the Historical Figure. *That* Jesus was all about nonviolence; he knew there was no way he was going to defeat the Roman Empire, yet he proceeded anyway. Also, and here Harold lights up, "Jesus Christ had a real theatrical flair, just as XR does."

Harold tells me a story that historians recently brought to light about Palm Sunday. Two thousand years ago, in the week leading up to Palm Sunday, Jews in Jerusalem celebrated the hoped-for overthrow and eviction of the Roman Empire. They were still under occupation then, and the Romans knew what they were celebrating and rightfully feared that the celebrations could give way to full-scale revolt. So each year in anticipation of that holiday, Rome dispatched a legion of soldiers to Jerusalem a week before the festivities began, to make sure nobody got out of line. The legion always arrived at the main gate of Jerusalem, led by a decorated general on a war horse in full regalia. There was no way to beat them, so Jesus decided to mock them, nonviolently. He rode a donkey up to Jerusalem's back gate, arriving just as the Roman soldiers did — only Jesus, instead of showing up resplendent on a towering stallion, wheezed up in shabby clothes on a tired old mule with his feet dragging on the ground, deliberately inglorious, every inch the clown, provoking laughter and cheers from his audience.

"That's what got Jesus arrested and crucified," Harold says, grinning. "Street theatre!"

We talk about the moral murk of today's struggle. There are no evil tyrants to overthrow; no oppressive legionnaires are forcing us to use internal combustion engines; the lowest person in society can publicly criticize the most powerful without fear of crucifixion.

"It's more a question of mass ignorance, increasingly aware and therefore culpable, but still far from deliberate malice," I say. "There's no moral comparison between overconsumption and murder."

"That's what makes the people whose minds we are trying to change worthy of our empathy," says Harold. "Like Jesus said when he was hanging on the cross, forgive them for they know not what they do."

Harold was very active in the politics of nuclear non-proliferation in the 1990s. But, like so many of the rebels I met, he'd only recently awakened to the consequences of human industry. I ask him why he chose to join Extinction Rebellion rather than Sierra Club or 350. org or any number of other environmental organizations who are always looking for volunteers.

He pauses to think, then laughs. "The honest answer is that I feel energized by catastrophe." He is a person who puts things off until the last minute, he says; only when enough pressure has accumulated does he spring into urgent action. Extinction Rebellion seemed to reflect that approach. No other group he came across seemed to feel the impending catastrophe of climate change as keenly as them. Everyone else was too cerebral. He met with people from the David Suzuki Foundation, for example, and they told him that their strategy was to focus on civic engagement; local politicians were the most sensitive to environmental campaigns. Harold appreciated their point but felt civic leaders were way too far down the power structure to enact the kind of sweeping, immediate changes the world needs. "I don't care if Vancouver increases its bus service," Harold says, shaking his head. "Maybe if we had until 2050 to get where we're going, but . . . It's too late for incremental change." And incremental change is all you'll ever get from working within the system, Harold says. The whole system is in need of sudden, massive change. That is Extinction Rebellion's message in a nutshell.

"I'm a little relieved to hear you talk about doubt," I tell Harold, "because I have some of my own with respect to Extinction Rebellion." Harold nods, unsurprised and unperturbed, as I confess how I'm drawn to the spirit of this movement but have a hard time with the institution. It comes across like it's the first group of people in the world to have noticed there's a problem. I've spoken to more than one rebel now who, when I asked why they chose to join Extinction Rebellion and not a different group, said they hadn't heard of any others. It was like they'd woken up yesterday, punched in a Google search for "environmental activist," and joined the first group that came up.

Harold smiles. He's used to hearing people say things they think no one else has said. "That's probably true," he says. "But maybe the world needs that fresh energy."

"Another thing I've noticed," I go on, "is that lately XR seems to be talking exclusively about climate change. Not just the Vancouver branch but the communiqués coming out of London, too. But isn't there a bigger picture? I guess I'm wondering if this was a conscious decision, which I can understand from a marketing perspective, or if everyone at XR is just sort of defaulting into the language of the moment."

"I honestly don't know," Harold says, "but my sense is that it's an unconscious collective decision. It hasn't come up at any of our meetings, but you're right — we do talk exclusively about climate change now." He thinks about it for a moment. "Maybe it's because the bigger thing is too complex and overwhelming to think about."

We both sit silent, thinking about the thing that is too complex and overwhelming to think about. My name for it is "growth." Harold's is "more." He tells me a story about his dad getting his first car, when Harold was four or five. "It didn't have a radio, and we didn't care! But could you imagine getting a car without a radio now? Nobody would accept such a thing. We've become so accustomed to always getting more of everything that it's come to seem normal. We've always got to have it. We're slaves to more. That's the thing that has to change. We've got to learn to love living with less."

I tell him I don't see overthrowing our government as a great solution, since the government of Canada strikes me as among the more representative democracies that exist on Earth. But I'm also torn, because so far our democracy has failed to make the necessary adjustments — it's all we can do to pass a piffling carbon tax. Does that mean democracy must go? Am I ready to say that? That seems to be what Extinction Rebellion is saying, but are they aware of it? Or, as with climate change versus growth, is this awareness just dimly percolating beneath the consciousness of XR members?

Harold doesn't know. What Harold knows is how it felt the other day when he took part in an XR training exercise downtown. He and a group of rebels walked into the street under a green light, all holding Extinction Rebellion signs, and stopped traffic. It was the most empowering feeling he'd had in years. It took him weeks to build up to it, and he was terrified going into it. Some people honked and were furious, and others were friendly and waved. In that moment, he felt like he was doing something.

"Do you think you were helping to awaken those drivers who got mad?"

"No."

"Then what was the point?"

"The point was to resist."

Was it selfish to block traffic just to make yourself feel better? Or was this a necessary gesture, the kind of tiny spark from which conflagrations grow? Harold's act of disobedience wasn't just one old guy blocking traffic for the thrill of it; it was an act of training that would flow into a larger act, with more people making more of an impact, in the hope that it would grow to encompass so many of us that it could overcome a system, or at least impact the system. That is a far-fetched hope — a terrible bet — but is it worse than doing nothing for the sake of not inconveniencing anyone?

False choice, you could say. Blocking or not blocking traffic are not the sum total of options available to us; there are countless ways to resist. But every one of them has its drawbacks. Every one of them will invite somebody's anger.

"I guess we should take hope from the resistance of Jesus," I say. "He took on the Romans and won against all odds."

"Oh, he didn't win," says Harold, laughing. "They slaughtered him and crushed the rebellion and kept ruling for centuries after that."

I seek outside opinions. One writer friend, wise and accomplished and with deep roots in environmental activism, says he thinks Extinction Rebellion is fantastic. "No great movement has ever succeeded without serious protest," he says. I tell him I worry they haven't thought things through, and he laughs. "You don't look to protest movements for depth of thought," he says. "You look to them for urgency." He notes that the obstacles to decarbonizing our economy no longer have anything to do with knowledge. The problem and its solutions are well understood, the path forward clear: some version of a Green New Deal, comprising massive investment in new energy infrastructure — electrifying everything with renewables, retrofitting homes, boosting public transit. It will be immensely strenuous to accomplish in an appropriate time frame but not all that complex — more like rolling a boulder up a hill than building a rocket. What's missing is the sense of collective urgency that motivates us to put our shoulders behind the boulder. "That's what groups like Extinction Rebellion are for," he says.

Another writer I know, equally smart and accomplished, is utterly disdainful of XR. She feels that they inadvertently set the cause backwards rather than forward. Canada, she notes, has come a long way on environmental issues; when Conservatives were in office from 2006 to 2015, we were essentially governed by a Calgary-based oil lobby. The Liberals may be flawed, but they have a strong progressive wing; they've championed women's and minority rights, led a global movement to phase out coal power, instituted a carbon tax at home, and can be pressured to go further on all counts. Extinction Rebellion, she feels, put all this at risk when they occupied Burrard Street Bridge and other bridges in major cities across Canada, just twelve days before a tight federal election. XR's antics were like a campaign commercial for a Conservative Party that had every chance of winning — although it's important to note that they didn't.

Curious what the environmental NGOs thought about the new kid on the block, I asked around. The informal consensus of grudging acceptance was captured by Sven Biggs, the Vancouver-based director of the oil and gas campaign at Stand.earth, a major international ENGO that has played a leading role in the fight to contain the oil sands. Stand.earth was among the organizers of a historic protest against Trans Mountain pipeline in 2018, which drew thousands of Vancouverites and saw over two hundred people arrested. "I think there's always a certain amount of tension between NGOs and grassroots activists," Biggs tells me. "We [the NGOs] have this responsibility to be extra careful with our resources because we're so outspent by the oil and gas lobby, and we have to be responsible to our donors, and often grassroots people are on the ground pouring their hearts and souls into something and they don't understand why Big Green doesn't show up and solve the problem. So there's an inherent tension there. I also think there's something unique with Extinction Rebellion — they're more of a movement than an organization. These kinds of movements explode onto the scene, same as Occupy, but how do you sustain that energy? Because it is all volunteer folks, and people have lives and other interests, and it's hard to keep focus on one campaign or one issue when there's so much going on in the world."

Biggs's observation that XR is a movement rather than an organization invites comparison with their famous American counterpart, the Sunrise Movement. Founded in 2017, Sunrise shares many broad similarities with Extinction Rebellion. Both were born of a desire to force government to act on climate change, a force applied by creative activism. Extinction Rebellion shot to greater and more immediate fame with its debut the following year, but Sunrise wasn't long in catching up, at least in America. (Sunrise is not an international group.) Among its greatest claims to fame is that Sunrise helped design the Green New Deal, through collaboration with politicians it helped get elected in the first place. The resulting impact on American climate policy has been extraordinary — the Biden administration's $2-trillion plan to make the entire U.S. economy carbon neutral by 2035 is essentially a Green New Deal in all but name.

But the Sunrise Movement is not a movement. It is a professional organization, an incorporated 501(c)(4) non-profit, to be exact — a political action group. That's the first major difference between it and XR. Sunrise may rely on a volunteer army, but it's run by paid professionals. You can apply for a job with Sunrise and draw a salary. Its volunteers are trained.

Its strategy is also radically different. Sunrise has focused not on overthrowing the government but on getting specific individuals elected into it. It juices the campaigns of climate-friendly politicians by creating viral campaign videos, alongside more traditional methods like phone banking and knocking on doors to get out the vote. It also does more traditional protest actions of the kind you can imagine XR doing. The 2018 sit-in of Nancy Pelosi's office (to pressure her endorsement of the Green New Deal) won Sunrise its first substantial press coverage in 2018; over 250 people joined that protest, including the newly elected Alexandria Ocasio-Cortez, and fifty-one were arrested. Since then, Sunrise has begun a "Wake-Up Campaign" where protesters gather to chant outside the private homes of climate-offending politicians at dawn — as XR-esque a stunt as I can imagine. But Sunrise's surgical focus on helping progressive politicians is what really sets it apart from just about any other climate action group on Earth. Of the twenty candidates they endorsed in the 2018 midterms, half won office. These included Alexandria Ocasio-Cortez and her fellow "squad" members Rashida Tlaib, Ilhan Omar, and Ayanna Pressley. Another candidate Sunrise endorsed was Deb Haaland; three years later, Haaland became the first Native American secretary of the interior and by far the most radically progressive.

Sunrise is radical, too, but it's also intensely pragmatic. From an Extinction Rebellion perspective, which spurns any and all compromise, the question might be: Is Sunrise radical enough?

↢∿↣

January 28, 2020

There's an action outside the head office of Teck Resources, a mining company awaiting permission to dig what would be the largest bitumen mine in Albertan oil sands history. The federal government is due to make a decision on the $20-billion Frontier mine next month.

Fifty or so rebels have gathered at the foot of the glass high-rise on Burrard Street; we're just a kilometre up from the bridge but this time no one's trying to block traffic. A wide concrete courtyard between the street and the entrance provides plenty of space for theatre. The Red Rebel Brigade, ten people draped in scarlet robes and veiled hoods to symbolize the common blood of humans and animals, silently stalks the courtyard; a Grim Reaper on stilts holds a sign: "Thank You Teck Mining. You Will Destroy the Environment Faster Than I Can." A few people hold up the trademark hourglass sign, someone is handing out hourglass pins, another person is handing out pamphlets that quantify the amount of carbon Teck's mine would release should it be built. The pamphlets don't mention that the project has the support of every First Nations and Métis community whose land would be directly affected.

The rest of us circle around the four horsemen of the apocalypse, who produce a basketball-sized globe and pour a jar of fake oil over it. Passersby on the downtown sidewalk spare a curious glance, some politely accepting a pamphlet without breaking their stride. Security guards make sure the building entrance remains accessible for employees in suits to come and go and otherwise let the proceedings continue. A young rebel sings to us through a portable speaker, then leads a round of chanted slogans, ending with the branded finale.

"What do we want?"

"Rebellion!"

"When do we want it?"

"Now!"

"Extinction!"

"Rebellion!"

No press shows up. Their absence is a popular topic of conversation, not just among Extinction Rebellion but throughout the activist community. The general consensus is that mainstream news outlets are ignoring the climate crisis in general and huge projects like this Frontier mine in particular. But that's not as true as it used to be, thanks in no small part to actions like this one and countless others organized by myriad groups within the environmentalist ecosystem. You could never look at any one street action or non-profit campaign and say, *That's the one that did it*, but their collective hum of outrage has decisively pierced the mainstream bubble.

The day after XR's Teck action, for instance, CBC Radio airs a highly critical twenty-one-minute documentary about Frontier, a subject such news outlets were supposedly ignoring. The sheer size of the project generates its own media attention; the breadth of environmental opposition has informed much of that attention, a fact not at all lost on the executives who seem to be ignoring us from thirty floors above.

Even as we're milling about on the pavement, they're getting ready to withdraw Teck's application for the Frontier mine in three weeks, before the federal government can make its decision one way or the other. "The nature of our business dictates that a vocal minority will almost inevitably oppose specific developments," Teck's CEO writes in an open letter explaining the decision. "We are prepared to face that sort of opposition. Frontier, however, has surfaced a broader debate over climate change and Canada's role in addressing it. It is our hope that withdrawing from the process will allow Canadians to shift to a larger and more positive discussion about the path forward." Translation: The usual rabble has managed to turn this project into a national shitstorm that isn't worth the headache. Pat yourselves on the back, tree huggers. You win.

But that victory is still invisible on this cold late January day. For now, the Red Rebel Brigade has nothing to celebrate. They weave their slow and mournful dance. Then the rebels begin packing up their gear to lead a march to the federal Ministry of Environment, two blocks away, where they will livestream their occupation of Environment Minister Jonathan Wilkinson's office.

I'd love to join, but I have to pick my daughter up from pre-school. No one seems surprised.

↞∿↠

By February 2020, Extinction Rebellion's founders in the U.K. have decided to narrow their focus, and the Vancouver chapter is following suit. Its next major day of action, in spring 2020, will not target the human condition or the global economic system. It will target the Trans Mountain pipeline project.

This is announced at the biggest General Assembly I've seen. Still a relative statement — over two hundred people have crammed into the church on a torrential weekday evening, but recruitment is humming and these rebels now represent a fraction of Vancouver's membership. Their energy is palpable. October's blockade of Burrard Bridge has already attained the narrative glow of a mythical founding event. Affinity groups (subchapters like Students for XR or Physicians for XR) are popping up everywhere and pursuing their own creative agendas; earlier this month, eight XR-UBC students went on a hundred-hour hunger strike that drew a promise from UBC president Santa Ono to divest the university's investment portfolio from fossil fuels.

There remains, however, the problem of loose cannons. Recently, a Vancouver Island chapter tried to stage a citizen's arrest of B.C. premier John Horgan. The rebels went to the premier's house and rang the doorbell; Horgan being at work, his wife answered and, upon seeing several large men shouting that she was under arrest, she called the police, who promptly came and conducted the other kind of arrest, which journalists showed up in time to observe. Those journalists got quotes from some rebels who'd hung back on the street, draped in hourglass banners. They explained the action was on behalf of the Wet'suwet'en Nation, then in negotiations with the B.C. and federal governments to resolve the Coastal GasLink pipeline dispute.

That conflict had inspired the biggest wave of protests this country has seen in a decade, with land protectors blocking vital train routes from British Columbia to Quebec, bleeding hundreds of millions of dollars from the national economy and threatening to deprive several municipalities of vital supplies, including chemical reagents necessary for keeping tap water potable. Into this delicate, nationally televised episode of Truth and Reconciliation blundered Vancouver Island's renegade (and very white) branch of Extinction Rebellion, stunting their citizen's arrest with all the finesse of Super Dave Osborne, and dedicating it to the Wet'suwet'en. The whole thing was supposed to be a metaphor, they tried to explain — trespassing on Horgan's property symbolized the state's trespass on Wet'suwet'en territory, and how did he like it when uninvited guests came knocking? — but the rebels hadn't consulted a soul in advance, and the Wet'suwet'en leadership immediately denounced the action for the damage it did to the negotiations they'd spent twenty years trying to secure.

"Sometimes we're just gonna step in the shit," says the speaker at Vancouver's General Assembly. She's the wiry, short-haired woman who often opens these meetings; I've come to think of her as Mother Fox. Right now she's explaining to the audience what happened at the premier's home and is walking a line between acknowledging the magnitude of the error (not at all self-evident to a good number of people in this pale audience) while leaving a path for redemption. "We're all new to this and we're learning as we go. It's gonna get uncomfortable. We're gonna make mistakes. Some will be big. When that happens, we have a choice. We can learn; we can own it and improve. Or we can give in to shame and give up."

THE ENVIRONMENTALIST'S DILEMMA

XR-Vancouver soon puts out a public statement that what happened at the premier's house was a mistake. This moment of soul-searching echoes a larger search taking place at Extinction Rebellion's global headquarters in London. The mothership has "unreservedly denounced" their most prominent co-founder, Roger Hallam, after he downplayed the severity of the Holocaust in an interview with the German weekly *Die Zeit*. Hallam's two-fold point was that tyrants have been pursuing genocide for millennia, and that climate change was set to kill far more than six million humans. Hallam claimed he was trying to get the public to see climate change in the same emotional terms as the Holocaust: as a moral crime whose cost in human life and misery make it a crime against humanity. A lot of environmentalists are sympathetic to that kind of argument. But a rich white man from England has to be careful when he talks about the Holocaust to Germans. He can't go calling the murder of six million Jews "just another fuckery in human history."

Hallam's disgrace, the uproar, and the cycle of apologies it provoked crystallized Extinction Rebellion's growing reputational issues. Chief among these is white privilege. The group can come off as more worried about koala bears than social justice; when human suffering does enter the equation, it's never today's victims of oppression that rebels invoke, only tomorrow's climate refugees. Many chapters also seem oblivious to the fact that many people of colour have a very different relationship with police and prisons than the middle-class white people in XR's core demographic, who often see arrest as something of a lark, a badge of honour, and hopefully someone takes a picture of the moment you get cuffed.

Another pressing issue is the movement's need to constantly one-up itself, pulling ever more outrageous stunts in order to stay in the limelight. When, in October 2019, one of the London groups (there are now dozens) blocked a commuter train at rush hour, a handful of rebels were pulled off its roof and beaten by an angry mob, and the whole movement got a black eye. The near-universal public response was to question not just the tactics but the logic of an environmental group disrupting the world's most environmentally friendly mode of transport.

Even as it's spread to over seventy countries, Extinction Rebellion is at risk of succumbing to Jonathan Matthew Smucker's political identity paradox. The stronger their core identity becomes, the crazier their stunts get, and the more they alienate the broader public.

The group's founders in the U.K. are well aware of this, and so are the people in Vancouver. No more mayhem for mayhem's sake. For the coming spring uprising, when rebels rise in tandem around the world, a more targeted approach is necessary, one that hamstrings an actual climate transgressor — not a bunch of moms and dads trying to get home in time to feed their kids.

The perfect target is right in our backyard: the Trans Mountain pipeline expansion project that Justin Trudeau's government bought a year ago for $4.5 billion. Construction on the project, which would twin the existing pipeline built in the 1950s — thereby doubling the amount of bitumen flowing across the province and tripling the amount of oil-tanker traffic going through the heart of this city — has barely begun. The pipeline's terminus in Burrard Inlet is a twenty-minute drive from downtown Vancouver. The project is hemorrhaging cash and political capital and looks highly vulnerable to mass action. Here's a federal government that can't stop talking about climate change and a new relationship with Indigenous Peoples, stomping over the bitter opposition of Vancouver's three coastal First Nations in order to plough through a pipeline for the express purpose of increasing oil production in Alberta. Vancouverites have already been protesting the pipeline on and off for two years — everyone from the city's current mayor to David Suzuki's grandson has been arrested — and with 3.5 million people living within a half-hour's drive of the tanker terminal, this could easily metastasize into the biggest act of civil diso—

Ah. Pandemic.

↞↝

August 26, 2020

It comes down to four of us sitting under hundred-year old cottonwoods at sunset.

We're three weeks into this, ah, occupation of the Brunette River watershed, just outside Vancouver. Even with the muffled roar of the Lougheed Highway's traffic filtering in through the canopy, it's easy to forget we're surrounded by city. The sky is turning orange-purple; the trees' white bark is, too; a hundred metres below us, ancient cedars lean over the bank of a small clear river that's been cleansed of last century's industrial poisoning and now awaits the fall salmon run that returned with its good health.

The Trans Mountain pipeline route runs directly underneath us. The company's been given a one-month window to clear these trees and lay the pipe before the salmon arrive in mid-September. After that, no one's allowed to touch anything till next summer, when the smolts have gone back to sea. And so instead of gathering a hundred thousand protesters under Extinction Rebellion's banner at the Burrard Inlet terminus, the city's few remaining rebels have joined a coalition of local environmental groups whose goal is to block this stretch of construction and set the whole project back by a year. The four of us (really three, since I don't count — I'm here to chat instead of help) are here as ground support for the person lying in a tent strung twenty metres up above us, in the canopy of the condemned cottonwoods.

COVID-19 dismantled Vancouver's branch of Extinction Rebellion, almost overnight. Many of its chief organizers left the city, some the country, to lay low with their families elsewhere. Those who remained went silent, preoccupied with all the things that preoccupied us once the world shut down.

The last person I saw before the lockdown was Lexa. We met at a coffee shop in the middle of March, and she described some of the growing pains that Extinction Rebellion was going through, as well as her own. The movement had taken every ounce of energy and free time Lexa had. After the better part of a year, she was burned out. She'd gotten her first job, in construction, through a friend in XR. I looked at her delicate hands, her fine-boned face, and tried to imagine what a construction site would be like for her. It was the last time I saw her in person. From then on, we spoke on the phone; she told me that she'd begun hormone therapy, that she was now called Lexa, and that she'd finally gotten an office job. She was living with her brother and his wife, but soon she'd have enough money to move into her own place. She still kept in touch with some friends at Extinction Rebellion, all of them supportive, but she'd decided to leave the group and no longer considered herself a member.

"XR filled a hole in my life," she told me. "I made a lot of friends there. I learned a lot and did my own research to see what direction we should head in, and I took a lot of satisfaction from that. I gained a lot of confidence. That was really good for me." But she'd given too much of herself to the movement. "It felt like it was kind of making me miserable, and I wasn't able to make progress in other areas of my life," she said.

At the end of her time with Extinction Rebellion, Lexa had pushed the group to evolve its strategy. "We need public sympathy on our side," she'd argued, "and in order to get anywhere we can't just go out and piss people off." She felt the public "just kind of saw us as radicals and didn't understand where we were coming from. And so as a result we had never really gotten numbers that we were looking for, like they managed to get in the U.K.

People would say, 'Well, we're not following the initial strategy from the U.K. closely enough, we need to be more confrontational, we need to be doing more extreme actions' — that kind of thing, getting more people arrested." But Lexa felt they should be moving in the opposite direction, pointing out that organizers in the U.K. were starting to say so, too. "What are we really achieving by getting more people arrested? Why do we need to be so confrontational, especially in a way that isn't necessarily strategic, blocking roads and public infrastructure? Is that really targeting the people we want to be targeting, to get political change? I think the consensus that we had gotten to as a group was, yeah, there's a lot of merit to this criticism; let's take it seriously and rethink our strategy a little bit. What would gain us more sympathy, rather than deteriorate our sympathy with the public?" But they never got the chance to find out.

The thing I remember most about our last face-to-face conversation, in a crowded coffee shop four days before the first lockdown, was how indifferent to the looming pandemic the people all around us were. We talked about that. Everyone knew the virus was coming, that some sort of lockdown was on its way in a matter of days, and yet the business of life went on without the slightest outward appearance of concern or change, as it did in the buzzing café where we sat speaking. "It's a pretty good metaphor, isn't it," Lexa said.

A few weeks later, at the height of the lockdown in mid-April, I spoke with Harold. He was surprised by XR's reaction to the pandemic. He'd been deeply involved in the organization, giving two or more talks a week and communicating every day with other organizers, treating it almost as a full-time job, and then suddenly everyone ghosted. Just like that. Nobody was communicating on Mattermost or any of the other private chat lines. Nobody said goodbye. His messages floated into the ether.

Not that he was looking to connect physically with anyone under the circumstances, and that was the real death blow. The pandemic, with its indefinite restriction on physical gatherings, could not have been more perfectly engineered to shut down an organization built entirely on mass gatherings. The U.K. headquarters wasn't quite as demolished; better financed and organized, it established an online presence that kept the group intact. But XR-Vancouver was so new and cashless and run entirely on upstart wingnut energy that COVID-19 dispersed them like a hot sun on fog. Their presence was reduced to sporadic, tone-challenged Facebook posts about the pandemic's positive environmental impacts.

As a result, spring construction on the Trans Mountain pipeline proceeded, out of sight and mind, still in its early stage and far from any cities. "Now is a great time to be building a pipeline because you can't have protests of more than fifteen people," said Alberta's perfectly named energy minister, Sonya Savage, a comment that managed to be accurate and outrageous at once. That was on May 22, three days before George Floyd was killed.

For the next two months, Black Lives Matter showed the world what it takes to get people in the streets. Somewhere between fifteen and twenty-five million Americans took part in those protests, a minimum of 4.5 percent of the population, shattering Chenoweth's 3.5 percent rule. To what effect? It would be facile to equate Donald Trump's ultimate defeat with the Black Lives Matter movement — one could as easily say they mobilized Republicans to vote as Democrats. But they weren't calling for Trump to be removed from office. They were calling for something almost as ephemeral as truth: justice.

The point is to resist. One tangible effect on XR-UK of the Black Lives Matter movement was that it issued a rare public apology in July: "We recognize now that our tactic of arrest has made it easier for people of privilege to participate and that our behaviours and attitudes fed into the system of white supremacy. We're sorry this recognition comes so late."

That didn't mean the rebellion was over, or even changed in any appreciable way. Two months after that apology, XR-UK returned to the streets in a series of actions that included blockading the printing presses of four national newspapers to prevent their distribution. More than ninety rebels were arrested over ten days; Britain's parliament, having acceded to their demand to declare a climate emergency two years earlier, now debated whether to declare them an organized crime group. XR-UK even got on the wrong side of David Attenborough. "We have to treat the people we share our community with with respect," he said on the BBC. "Disturbing their lives to such an extent that innocent people can't go about their own business is a serious thing to do and could disenchant an awful lot of people."

But in Vancouver, a version of Extinction Rebellion that Attenborough might have readily endorsed began to reassemble itself that summer. In mid-July, with the pandemic at a low ebb, I started getting emails again. In early August, the rebels called an outdoor meeting at Hume Park, right beside the Brunette River watershed, to discuss that month's tactical strike against Trans Mountain. This time XR-Vancouver wasn't acting alone but in coordination with a number of environmental organizations and First Nations, all of whom had been fighting the pipeline project for years.

At the beginning of August, two days before the protest began, we met underneath an ancient and enormous weeping willow in Hume Park. Fifty people showed up, including some familiar faces behind the masks — there was the Fox Mother, there was the PhD, there was the Red Rebel Brigade. Harold, thirty years older than the rest of us, was out. Lexa, living her own life for the first time in her life, was out.

This time, there were some First Nations people who had been invited to speak to the group. A Secwepemc woman with her teenaged daughter sitting shyly at her feet told us about this struggle's importance to her people; she belonged to a group called the Tiny House Warriors, Indigenous land protectors who'd built and moved into small houses along the proposed pipeline route, many of whom had suffered violent attacks from white pro-pipeline neighbours. Far from deploring Extinction Rebellion's white privilege, she reminded us we had an opportunity to put it to good use in the weeks ahead. She told us about the vile treatment she'd suffered at the hands of police and asked us to bear that in mind when and if we confronted the police ourselves.

And there was Tim Takaro, a sixty-three-year-old physician and professor of environmental health at Simon Fraser University, who led the group on a tour of the watershed we were about to protect. "This is the right struggle at the right time," he said. What he didn't say was that two days later, he'd be the first person to ascend a rope and spend a week living in a tent amid the treetops.

The month-long occupation of the Brunette River watershed that followed was the most anticlimactic, news-killing story I've ever witnessed. It consisted of one person hanging up in the trees for a week, then another person, then another person, and then another. No police or construction crews bothered to show up. The newspapers — much to everyone's consternation — took no more notice of Dr. Takaro and the tree-dwellers who replaced him than did the salmon when they finally arrived on the scene.

Nothing I wrote was going to help them now. It was unlikely to help them later. But no one seemed to mind when I popped by periodically over the following month. I'd stroll into the trees, enjoy the reminder that wilderness can squeeze itself into the tightest of man-made spaces, and chat with whoever happened to be among the 24/7 ground support crew.

My whole previous year with XR was distilled into one late August sunset when I sat with three others minding the camp. Two were rebels, a crisply dressed young man and a kind woman in her late thirties with Naomi Klein's latest book in her lap; the third was a Forest Church pastor, a fifty-something mother whose children were still at home but old enough now that she could re-engage in the environmental activism her faith had called her to all her life.

The days were noticeably shorter now, but the full warmth of summer still infused the gathering dusk. For a while, we weren't rebels or writers or pastors or parents, just four people with a common concern trying not to let whatever might come next prevent us from enjoying a calm evening in the woods.

The conversation turned to hope. I said I believed it could be a radical response to dire circumstances. The rebels felt it was more of an opiate for the masses, the kind our prime minister loved to administer. The pastor was of two minds.

"It's when hope is most unreasonable that it's most powerful," she said, offering as an example her own belief that death was a beginning and not an end. "But lately," she added, gesturing at the trees around us to indicate our common cause, "I've been feeling more bloody-minded than hopeful."

I'd seen the young man, a university student whose parents were from Pakistan, every time I came around that August; with plenty of time on his hands between semesters, he was stepping into the leadership vacuum the pandemic had created in XR-Vancouver. I asked him how he felt about the perception some people had that Extinction Rebellion was a very white movement. "I read those critiques," he said, "and it felt to me like if too many people think that's the way it is, then that becomes the way it stays. I've never had any trouble with it here." He wasn't disagreeing that mistakes had been made, in Vancouver and elsewhere, but he felt like he was among family here, had always been accepted, thought maybe the prominence of Coastal First Nations in Vancouver's environmental scene made this chapter of XR more sensitive to racial dynamics than was the case in other cities and countries.

His comments brought me back to a brief encounter two weeks earlier, during that summer's first and only XR street march. A hundred people had gathered by the willow tree at Hume Park to mark the start of August's fight against Trans Mountain. As the crowd set forth to block the nearest major intersection, I saw someone I recognized but hadn't ever met: Will George, a Tsleil-Waututh activist well known around here for his public role in the fight against Trans Mountain. Will George belonged to a prominent Tsleil-Waututh family of leaders and activists and was arguably the most radical; he'd helped organize an action that saw several people suspend themselves from the Second Narrows Bridge to hang above oil tankers passing below, and he'd interrupted more than one of Trudeau's press conferences when the prime minister visited B.C. George cut an imposing figure, powerfully built with wraparound shades and tattoos that licked up from his shoulders to wind around his neck. When I introduced myself, he gave me a look that was simultaneously friendly and wary. But he was happy to talk.

I asked him how he felt about Extinction Rebellion. Had the group been a sound ally to the Tsleil-Waututh Nation in its decade-long struggle against this pipeline?

"They're good," he said. "We've had to teach them a few things. When they first started coming around, they always wanted to put their stamp on everything." Extinction Rebellion's branding — the stylized hourglass, the monogrammed banners, the self-referring chants — was not a great asset on First Nations turf. "But we told them to cool it and they listened. They're willing to put their bodies on the line for this fight. We appreciate that."

Fleeting as the interaction was, it added to my sense that Vancouver's chapter of Extinction Rebellion was less radical and more responsible than its counterparts in London. It also made fewer headlines. None of the rebels had been arrested in the year since the Burrard Bridge action, and they weren't planning anything special for September, a month when XR-UK once again logged dozens of arrests and pissed off their whole country.

As summer gave way to fall, the occupation of the Brunette River watershed dragged on unnoticed. No one sent a reporter when the tent in the canopy became a permanent wooden tree-house. Instead, the pandemic's second wave and the tumult of American politics served to mute the significance of a few bedraggled land protectors watching over a rain-soaked patch of earth on the edge of a political backwater. One day shortly before Christmas, the RCMP finally moved in on behalf of Trans Mountain and cleared the land protectors out of their camp. Nobody noticed that, either. Then a few weeks later, my wife and I were driving home from the North Shore and found our usual route past the port of Vancouver closed off by squad cars, rerouting us into a slog of similarly diverted eighteen-wheelers. That evening when I checked my phone, I learned the traffic jam was caused by Extinction Rebellion. They'd blockaded the port to raise awareness of the Trans Mountain pipeline expansion project, racking up the first handful of arrests for 2021.

What was the point? *To resist.* But what was the point of resisting? The only people who could have learned about the action were those who already followed XR-Vancouver on social media. The executives and politicians pushing Trans Mountain probably got word, but I doubt they were any more inconvenienced by it than we were. The rebellion had lost its momentum, both here and abroad. Its energy was spent.

Or had it just dispersed? The world had changed enormously in the brief time since Extinction Rebellion exploded on the scene in 2018. The U.S. and China were now locked in a race to decarbonize, the world's automakers had kicked off a parallel race to electrify their fleets, and climate change was finally a top-tier election issue in almost all the world's advanced economies. Extinction Rebellion couldn't take credit for all this, but its arrival at a key historical juncture had been one of many defibrillators shocking the body politic.

Even if the global economy has upgraded its climate targets, we're going to need more defibrillators. The worst of the ecological crisis is still ahead. The pressure to ignore it, or downplay it, or delay acting on it, or profit from it will continue to rise. As the embers of resistance smoulder on amid the remains of groups like Extinction Rebellion, maybe the right question to ask isn't why resist, but rather: Where will the next spark catch?

INSIDE JOB

October 26, 2020

HEATHER MCPHERSON IS HAVING A DAY.

Her first caucus meeting was at six in the morning. That's fine. Now the Member of Parliament for Edmonton Strathcona has been in the House of Commons for twelve hours and counting. Also fine. McPherson's New Democrat Party, holding the balance of power in Canada's minority government and thus caught downstream of the latest pissing match between Liberals and Conservatives, has had to decide for the second time in a week whether the Liberals' latest fuck-up merits a vote of non-confidence, which would trigger a federal election, which if current polling is remotely accurate, would only hand the Liberals the majority government they crave. Less fine. For the second time in a week, McPherson and the rest of the NDP voted to keep the government alive.

Sparing Canadians a fall election also means cancelling the political colonoscopy that Conservatives so badly wanted to perform on Justin Trudeau: an ethics investigation into the $912-million contract the federal government awarded the Toronto-based WE Charity in spring 2020. That contract would have created a few thousand summer jobs for students — on paper, a nice little win-win for the Liberals, simultaneously meeting the NDP's demand to rescue cash-strapped students while enlisting more workers in the fight against the pandemic. Alas, as everyone in Canada immediately learned and will soon forget, the charity in question had recently paid the prime minister's mother, brother, and wife for speaking gigs and also happened to have employed both the finance minister's daughters — a revelation that caused the Conservatives more joy than any amount of student assistance ever could. The contract was cancelled, the $912 million went unspent, no students were helped, the charity itself imploded under all the scrutiny, and here we are still talking about it while COVID-19 begins its second siege of human civilization. The whole thing was so typically Canadian, the kind of drama you'd expect to read in a small-town paper ("Mayor hires neighbour's kid to build a lemonade stand, changes mind, nobody gets lemonade"), but instead it's made national headlines and exasperated forty million citizens who have bigger deals to worry about, Heather McPherson very much included, for four months.

And since McPherson and her party have just denied Conservatives their colonoscopy, the only dopamine hit the party of fiscal restraint could salvage from this debacle was to call the NDP "the Liberals' lapdog" and watch journalists chase that little bone around the internet for a few hours.

Now it's past seven p.m. and Heather McPherson is still on Parliament Hill, standing in a hall outside the chamber and talking on the phone to a pesky journalist from Vancouver who's asked her why the NDP hasn't used its leverage in this minority government to push for more climate action.

"Because I literally have had no one contact my office about climate change since March," she says. "Because right now [Premier]

Jason Kenney is decimating Alberta health care; he's inching closer and closer to this privatization line and the federal government isn't doing anything. So I'm going to stand in the House in half an hour to talk about a motion I'm putting forward to reaffirm a commitment to the Canada Health Act."

Just this morning, Alberta's health-care workers, aghast at Premier Kenney's decision to turn eleven thousand health-care jobs over to private contractors, launched a wildcat strike that halted all non-essential surgeries across the province to protest his cuts, "and the strikes are going to go on, and it's going to spread and there's probably going to be a general strike in Alberta. And so all of a sudden this has to be the thing that I'm going to talk about. But it isn't the big picture. It matters an awful lot, but where's the space to have these bigger conversations? I met with Amnesty International today and we talked about human rights abuses that are happening around the world. Well, it's really hard to find the space for that. I pitched a question today on nuclear disarmament because it just became international law on Saturday and that's a big deal. We can't talk about it because there's no water and they're evacuating Indigenous communities right now."

Neskantaga First Nation in Northern Ontario ran out of drinking water a few days ago — not tap water, which they've never had, but *any* water — after an oil slick appeared on the reservoir from which the community draws its buckets, so now the whole community has been evacuated to Thunder Bay.

Another subject that sucked up parliamentary oxygen today was conversion therapy. "That we're going to have legislation that says you can't have conversion therapy is great, that's fantastic, but at no point today did we talk about climate change," McPherson says. "At no point did we talk about workers in Alberta. At no point today did we talk about the massive things that are facing us."

Still, I reply, the NDP has the power to bring this government down. That's a hell of a bargaining chip. They've used it repeatedly in the year since McPherson took office, to push the Liberals to dramatically raise federal support for students, unemployed workers, small businesses, child care, people who are sick or have disabilities. McPherson and her colleagues have also talked a great deal about the opportunity this pandemic has created to rebuild an environmentally sustainable economy; why haven't they pushed for it in parliament?

"Right now, people are so overwhelmed with their own lives that they're forgetting this bigger-picture piece." It's not their fault, she adds, and now the stories tumble out of her just as they must have tumbled in. "I've gotten calls from sixty-year-old men who are bawling. They were so clearly told that if they worked hard all their lives — these were oil and gas guys, they certainly weren't people who voted for me — but we sold them this myth that if they worked hard they'd be okay. They were unemployed *before* COVID hit, and they had nowhere to turn. What do you tell them? There's nothing coming back for you right now. I'm getting calls from teachers who are weeping — their job was hard before, and now their job is to keep themselves and their families and their loved ones safe *and* keep my kids safe, and keep your kids safe. It's overwhelming and they don't have the tools to do that. They're not doing well. So they're not thinking about the climate crisis at the moment."

She pauses for a moment, gathers herself, thanks the person who just brought her an espresso on the other end of the line, and apologizes for sounding negative. We've been having these occasional conversations since she was elected one year ago, almost to the day. A year of best laid plans.

She sighs, allows herself a wry laugh. "I'm not in a very good state, I'm realizing."

The first time I heard the name Heather McPherson was October 27, 2019, when the Conservative Party swept all but one of Alberta's thirty-four parliamentary ridings in Canada's federal election. They took all of Saskatchewan, too, and most of Manitoba, leaving just one orange dot in a prairie sea of blue. That dot was the south Edmonton district of Strathcona, where McPherson grew up, and so did I.

McPherson was brand new to politics, a transplant from the NGO sector where she'd dedicated most of her life to international development. Her career began overseas, working for organizations like Médecins Sans Frontières and Canada World Youth on contracts that took her throughout Africa and Latin America. By the time she ran for office, she'd been executive director of the Edmonton-based Alberta Council for Global Cooperation for ten years. The ACGC is an umbrella group dedicated to promoting the UN's sustainable development goals among Albertan civil society; a big part of her job there was pressuring governments — her own as well as others — to improve the way they dealt with human and environmental rights. Now she'd jumped the fence. I wondered why, what she thought she might accomplish, how that would measure up to her experience. I wondered how it felt to be the only parliamentarian from the fourth-largest oil and gas producing region on Earth who gives an honest damn about climate change.

Not that she's the only lefty in town. Strathcona, which encompasses the University of Alberta and its arts and theatre district, is a deeply progressive neighbourhood. For a long time, it was sufficiently distinct from the rest of Edmonton to be affectionately known as the Gaza Strip. McPherson had a harder time winning her primary against a far-left environmentalist than she did winning the general; she defeated her Conservative opponent by a landslide. But Strathcona's ethos has been percolating through the city that surrounds it for a long time now. Edmonton's intellectuals and artists and filmmakers have crossed the North Saskatchewan River and spread throughout the city, and the Ledge — the domed provincial legislature that rises on the North Saskatchewan's far bank, controlled by Conservatives for all but four of the last fifty years — has come to be surrounded. Out of Edmonton's twenty legislative districts, eighteen are held by the provincial NDP.

That said, this is still the city where a street mural of Greta Thunberg can't make it two days before being defaced. "Stop the Lies. This Is Oil Country!" was scrawled over her face in black spray paint immediately after Thunberg's visit to the Ledge in October 2019. Which maybe helps explain why the provincial NDP's success here hasn't translated into federal NDP votes, with the sole exception of Edmonton Strathcona.

So it's fair to say that the district we grew up in remains the progressive heart of a city that likes to call itself the biggest small town in Canada.

That started to feel true when I began asking around about McPherson. Some of my friends knew her from the advocacy world. It turned out we'd both gone to Strathcona High School, though she graduated two years before I began. When I saw her picture twenty-five years later in an article about the district's new MP, I noticed something familiar about her smile, the way it balanced on the edge of a devilish grin. Then I realized I knew her younger brother. He and I had been friends in high school. I hadn't seen him in two decades, but I still remembered that grin.

Any Canadian can tell you that Edmonton is flat and cold and isolated, a three-hour drive from the nearest city (Calgary) and farther still from the Rocky Mountains. Between November and March, the temperature regularly drops under thirty degrees below zero, but you don't get to complain until it reaches minus forty. All my life, I've received sympathetic looks when I mention where I grew up. Naturally, that's only strengthened my love for the place where I learned how to ride a bike, catch butterflies, kiss, roll a joint, calculate derivatives, and appreciate Eddie Vedder.

What every Canadian cannot tell you, but is understood by the people who stayed or moved here, is that Edmonton is a city that remains affordable enough for an artist or an activist to live in without needing three jobs and four roommates. It's a city that supports creative risk-takers in ways that Vancouver or Toronto are less inclined to do, a city that's never tried to emulate New York and has always felt free to be itself — welcoming, blue collar, aware of and indifferent to the world's superiority complex, a collegial underdog that hasn't been noticed since Gretzky moved to California, with a swiftly growing population of immigrants to aid in the development of different ways to be.

The reason Edmonton has managed to be an urban version of an indie filmmaker is, of course, the money that fossil fuels have provided. That money started pouring in shortly before I was born, and it has flowed all my life, manifest in the quality of Alberta's schools and hospitals, its infrastructure, its status as the only province in the country not to have a sales tax and the cheapest gasoline in Canada.

The first phase of Alberta's oil wealth was surprisingly well managed. Or rather created — the industry required a great deal of government investment to get off the ground. The primary investment vehicle was an oil company called Petro-Canada. Founded in 1975 with $1.5 billion in taxpayer money, Petro-Canada laid the infrastructural groundwork for the free-market bonanza to come. It was voted into existence by Pierre Trudeau's minority Liberals, with the support of the federal NDP, over the objections of Conservatives in Ottawa and Edmonton.

Despite his distrust of free-market interference, Alberta's premier at the time, Peter Lougheed, had the foresight to establish the Heritage Savings Trust Fund that put away a portion of the profits in an endowment — the first and final acknowledgement by Alberta's ruling class that oil wouldn't always provide the province with a third of its income. When I graduated high school in 1994, a $1,500 Heritage grant for post-secondary education (enough for a year's tuition) awaited me.

Petro-Canada was turning a profit by then and no longer publicly owned; the Conservatives privatized it as soon as they'd taken back Ottawa. The sharp and lawyerly Lougheed was long gone, too. In his place was a boisterous cowboy named Ralph Klein, a high school dropout with a drinker's ruddy nose and the charm of a Texas barbecue. Klein became premier in 1992, just as I was starting to pay attention to the world beyond my neighbourhood, and he governed Alberta for fourteen of its goldenest years. Depending on their place on the political spectrum, Albertans either remember Klein jovially flipping pancakes at the Calgary Stampede and cracking jokes under a broad-rimmed cowboy hat, or barging into a homeless shelter one night, drunk and dishevelled, to berate the residents for not having a job. He threw a handful of coins at them on his way out, an act he reprised on sober second thought in 2006, when his government sent a $400 cheque to three million Albertans in celebration of the booming budget Alberta's oil sands had bequeathed — Ralph Bucks, as they were known. One year later, the global financial meltdown set Alberta on a new trajectory. There has been more than one mini-boom since, and Alberta's kept pace as the wealthiest province in the country, but the writing on the wall has been unambiguous for quite some time. The best you could say of the province's unwavering commitment to oil and gas is that it's made for some volatile budgeting.

That volatility hit the Conservatives hard. After Klein's protracted reign, they turned in on themselves, burning through four leaders in eight years and fracturing into two parties, a split that handed victory to the provincial NDP in 2015. The first Conservative loss in living memory was a seismic event in a province where Conservative identity is as deeply and famously rooted as Republicanism is for Texas. The shock was magnified six months later in the federal election, when the climate-preaching Justin Trudeau put an end to ten years of Conservative government under Stephen Harper, the Calgary economist who embodied oil's grip on the imagination of the governing class in Edmonton and Ottawa alike.

Alberta's Conservatives, at least, recovered quickly. After 2015, stewing in opposition, they merged back into a single party and elected one of Stephen Harper's former cabinet ministers to lead them. The result was inevitable. In 2019, half a year before Heather McPherson joined the federal fray, the new United Conservative Party (UCP) swept easily back into provincial power with Jason Kenney at the helm.

But four years in the shadows, along with a pronounced global shift in the investment climate, all on top of a remarkably successful international campaign to ostracize the oil sands, had changed the tenor of Alberta's ruling Conservatives. Gone was the cheerfully drunk Klein and his easygoing colleagues, throwing money at the people while the royalties flowed in. Here was the seriously sober Jason Kenney: a man who knew how to wear a cowboy hat and could fit right in with the good old boys, sort of, if only time allowed. Instead, he was constantly besieged by pipeline protesters, unions, and the public sector. Abandoned even by the price of oil, Kenney returned the attack, outlawed every enemy he could and became ritually angry at a Prime Minister intent on stealing Alberta's oil wealth, just like his daddy had. Never mind that Trudeau Junior's support for the Trans Mountain pipeline expansion project torched Liberal prospects in B.C. for a generation — according to Kenney, Justin Trudeau hated Alberta and deserved to be hated back. It was

Justin Trudeau's job-killing carbon tax and pipeline-killing regulations, directly descended from Pierre Trudeau's National Energy Program, that were to blame for Alberta's cratering economy. How else to explain it? Billions of barrels of bitumen languishing in the ground, civilization burning through record amounts of oil, yet Alberta was earning more from cannabis and liquor sales than it was from oil and gas.

The impact on Conservative psychology was something to behold.

The United Conservative Party cancelled Trudeau's federally imposed carbon tax (a move the Supreme Court later judged illegal), then saw no irony in being forced to cancel the subsequent celebration because of all the smoke from unseasonal forest fires.

It was the party that offered a 10 percent tax cut to corporations, only to see the biggest ones — Husky, Cenovus, Suncor, and other fossil fuel behemoths — lay off a quarter of their workforce.

The party that confidently based its 2020 budget on oil prices of $58 per barrel, then watched the actual price stay below $40 for the duration of the year, dipping all the way below zero for a few memorable days in April.

The party that invested $1.5 billion in the Keystone XL pipeline, on top of guaranteeing a $6-billion loan for the project, while Joe Biden campaigned on a promise to cancel that same pipeline.

The party that made protesting energy infrastructure a criminal offence, then opened 53,000 square kilometres of Rocky Mountain foothills to open-pit coal mining (overturning a moratorium that had been in place since 1976).

The party that created an information "war room," with a $30-million annual budget, to battle anti–oil sands propaganda with pro–oil sands propaganda, an effort whose primary method of attracting notice was to publicly display a spectacular incompetence. Within days of the war room's official launch, it came out that its logo was the fruit of plagiarism, pilfered from an

American software firm with the exact same graphic; a few days after that the war room's propagandists were caught impersonating journalists in order to land interviews with unsuspecting Canadians who might provide nice quotes about oil and gas; a month later, the war room was forced to apologize to the *New York Times* for accusing the Gray Lady of anti-Semitism. These slapstick own-goals set the tone for an ongoing comedy of errors whose chief accomplishment was to provide critics of both the UCP and the oil sands with a steady stream of material.

There's a new adjective for this combination of bumbling inanity and malignant devotion to lost cause: Trumpian. Kenney's UCP serves as a useful reminder that Canada is not immune to the illness that brought the United States so low. These things are contagious, and the primary vehicle for their contagion is the Conservative party, both federally and in Alberta. In the fall of 2020, an Angus Reid national poll found that support for Donald Trump among Liberal and NDP voters stood at 2 and 3 percent, respectively; among Conservatives, it was 41 percent. This is most pronounced in Alberta, where Trump was more popular by the end of his term than in any other province. A 338Canada poll conducted in October 2020 found that Trump enjoyed a 32 percent approval rating in my natal province — not exactly popular, but still twice the Canadian average and almost double the next highest region.

It wasn't just the reckless commitment to fossil fuels that reeked of Make Alberta Great Again, but the character tics and traits that came with it. Like the Trump administration they palely resembled, the UCP followed one form of denial into a labyrinth of others. Take the war room. A government at war with the fact of climate change creates a propaganda machine that purports to fight fake news with its own stream of misinformation.

Take the emphasis on (certain) rights and freedoms as a reason to avoid COVID lockdowns and mask mandates, even as Alberta led the country in new and active cases. Forcing bars and casinos to close, Kenney insisted, would be "an unprecedented violation of fundamental, constitutionally protected rights and freedoms."

Take the romantic idealization of Canada's colonial history, captured perfectly by the UCP's very public outrage when protesters toppled a statue of John A. MacDonald in Montreal, and Kenney responded by asking Montreal to send the decapitated statue to Edmonton so that he could erect it in front of the Ledge. (They didn't.)

Take Kenney's minister of agriculture and forestry, Devin Dreeshen, who spent much of 2015 in the United States volunteering for the Trump campaign. Take Kenney's speech writer, a former journalist who has published statements like "everyone knows that race is the defining element of violent crime in Canada," and called the residential school legacy "a bogus genocide story." The UCP is the party that considered removing the history of residential schools from the elementary school curriculum, alongside other changes such as teaching kids that most non-white Albertans are Christians. This is the party whose education minister's chief of staff was caught posing for photographs with the white supremacist group Soldiers of Odin. The party whose member from Lac Ste. Anne publicly complained that federal COVID relief cheques were "funny money" being spent on drugs and cheezies by recipients who would rather watch cartoons than get a job — this at a time when the province's unemployment rate had nearly doubled to 12 percent, higher than anywhere in Canada except the Maritimes.

Call it a hostile takeover, call it a Petro State, call it mob rule that's finally turned in on itself. An Albertan way to say it is the province has gone off the rails. You could almost find it funny if you didn't know anyone who lived there. But really, you'd be laughing at yourself, because what happened to Alberta is just a concentrated form of what happened to the world.

Heather McPherson has always had a strong sense of fair play, but it wasn't till she spent six months in Zimbabwe and Mozambique after finishing her undergrad that she got a close view of what unfair really looks like. "That was the first time I saw inequality on that level, the first time I had friends who were unable to meet their basic needs," she told me on a wintry afternoon in Edmonton, just before COVID struck. "That's the one where you start to think, *These aren't African people or Mozambicans — this is Juma. My friend. These are my people.*" She planned to stay longer but she was violently mugged with a beer bottle.

McPherson flew home to convalesce with several stitches and a shaved head, and a new determination of how to spend her life. She returned to Africa, this time on contract with an obscure NGO to help local governments in Uganda develop new computer systems. "I was this eager young kid that really didn't have a clue what she was doing," she said, rolling her eyes at the memory of her first experience with the white saviour industrial complex. "They put me in a community that didn't even have electricity, let alone computers." Disillusioned but undeterred, McPherson went on to volunteer for Médecins Sans Frontières for two years, then spent four more working for Canada World Youth in a number of communities throughout sub-Saharan Africa and Latin America. "I learned that I was pretty good at building coalitions," she said. Along the way, she completed a master's in international education and wound up at the head of the Alberta Council for Global Cooperation. McPherson held that position for the next decade, gathering an intimate familiarity with progressive institutions and politicians from Alberta to Ethiopia.

She's the kind of person who can say things like, "I want every Albertan to think about their role in the world and what it means to be a global citizen," and make it sound like a mischievous caper

— so corny and earnest that it's counterintuitively badass. A *let's have another drink and plot the betterment of the world* kind of vibe. We were in fact having a beer at the Next Act, three blocks from her constituency office. A few patrons came by to say hello over the course of our conversation, which lasted an hour before she had to pick up her son from school and take him to hockey.

When McPherson's predecessor, Linda Duncan, first approached her to run for the NDP in the 2015 election, McPherson's two children were six and nine years old — too young, she felt, for their mother to be away so much; an MP typically spends more time in Ottawa than at home. McPherson said no. Until then, the thought of entering politics had never crossed her mind. After Linda Duncan's overture, she started thinking "that maybe this was an opportunity, but also maybe an obligation."

That was the year the United Nations established its 2030 Agenda for Sustainable Development, which laid out a blueprint for each of the UN's member states to pursue seventeen sustainable development goals. "Sustainable development" has come under a lot of fire over the years for being an empty buzzword at best, an oxymoron at worst. But to McPherson, the sustainable development goals articulated a sophisticated new approach to a suite of age-old problems. The SDGs explicitly linked issues like poverty, education, and gender equality to climate change and justice. "I'd always been frustrated by how the development community viewed these as separate issues," McPherson said. She'd witnessed how the need to walk two miles for potable water could hamstring a child's education or compromise a woman's safety. "The SDG framework explicitly de-siloed these things. That was a major step forward." McPherson immediately put the SDGs at the heart of the Alberta Council for Global Cooperation's mission statement, talking them up every chance she got.

In the fall of 2016, McPherson, an avid runner, started noticing her jogs were leaving her more exhausted than usual. "I thought maybe this is just what old feels like." She mentioned it in passing to her family doctor on a routine checkup, and the doctor took some blood samples to check her iron levels. Two weeks later, she was diagnosed with colon cancer.

One month after her diagnosis, McPherson had sixteen inches of her colon removed. The surgeon got all the cancer on the first shot. "If you're ever going to get cancer, you really want to do it this way," she said. "They caught it fast enough, they got it out of me, and that was that."

Her brush with mortality had the clarifying impact you'd expect. "It brings into stark relief that you are not immortal and you are running out of time," McPherson said. She was back at work by spring, and when Linda Duncan approached her once more to run in the next election, McPherson agreed. Her kids were older now, and having survived cancer "went a long ways toward me saying yes." It also fed her fury three years later, when Jason Kenney began dismantling the public health-care system that had saved her life, a point she frequently makes in the House of Commons.

I asked her what she hoped to accomplish as a politician that she couldn't in her previous life. "Well," she said, "you've known my brother for twenty years but really didn't want to talk to me at all until I became an MP." She was joking, sort of, but the point was clear: She had a platform now that she'd never had before. A voice in the national conversation.

It wasn't just that. "There are a lot of members of parliament who really like being members of parliament and don't think there's much wrong with the system. My whole career has been working in systems that shouldn't exist the way they do. International development should not be a multibillion dollar industry! That's the whole point of this industry, that it shouldn't exist. Same with environmental activism. The whole point of an activist is to make it so that there's no need for an activist. And so some of the systems that we've got within parliament need to be changed. People need to get down and do this work. I'm one of those people."

McPherson was named the NDP's critic for international development and the deputy critic for foreign affairs (once again, she noted, the unnecessary siloing of clearly connected departments reared its head). A big part of her learning curve was figuring out how to balance the local demands of her constituency with her national duties. "But to be honest, right now my constituency is trumping — ah!" She gave a laugh and had a sip as if to wash her mouth out. "I can't even use that word anymore.

"But Alberta is in such a state. It's devastating what's happening right now. It's not even just the economic slump, it is the visceral attacks on our identity. It's all of this scary Wexit separatism and dog-whistle politics. The narrative is so ugly here right now. You have gaslighting politicians saying that nurses don't do their job. You have teachers in Alberta and doctors and specialists saying that they don't want to work here anymore. It's this carpet bomb of our province, and the labour movement is baffled, the teachers are baffled, the entire province feels like it's under attack. And I know it's partly my echo chamber, but even outside of my echo chamber, Albertans are feeling under attack. You've got this oil and gas sector that *is* under attack. Because oil and gas prices are not going forward, because oil and gas companies are automating, and because the climate crisis means that their product isn't as necessary, or can't be as necessary. So they're totally afraid, starting to feel very hopeless, and all of this fear and all of this anger is boiling up into unlivable racism. You know, anti-Indigenous sentiment that is absurd and absurdly dangerous. And it feels to me like there are politicians that are just dropping matches everywhere.

"We have to remember that we weren't always oil and gas. We got this oil and gas gift, and we should be proud of it, but it's been decades that we've known this is coming and our government hasn't done anything. Our government has in fact made it worse by doubling down, by turning up the hyperbole on us versus them. There's been no political discourse at all about what is the common sense thing for us to do next. Nobody wants to talk about that. Even the provincial NDP that was in power before . . ." Here McPherson trailed off. She'd entered dangerous territory.

Just before she took office, Alberta's NDP tasted power for the first time since it formed in 1932. From 2015 to 2019, they ruled the Ledge under Premier Rachel Notley. Notley took climate change far more seriously than any Conservative premier ever had, targeting the worst of Alberta's three fossil fuels for elimination. Under her watch, Alberta instituted a plan to phase out coal power by 2030 — no small task for a province that relied on coal to generate more than 60 percent of its electricity. Over the next five years alone, the province lowered its coal footprint by half and remains on track to eliminate it entirely by 2030. There is no other comparable source of emissions reductions in Canada.

Notley also provided crucial support for Justin Trudeau's national climate plan in the wake of the Paris Agreement, accepting the imposition of a national carbon tax that Kenney later tried to repeal. But Notley stopped short of imposing any meaningful restrictions on oil and gas expansion. In exchange for Notley's support, Justin Trudeau promised to back the Trans Mountain pipeline expansion project, which was vital to Notley's re-election strategy. Their compact infuriated the environmental and many First Nations communities, who felt deeply betrayed by both leaders. Notley, in turn, felt betrayed by the federal NDP, which came out against any and all new pipelines, period. Notley's take on that position yielded one of her most famous quotes: "Here in Alberta we ride horses, not unicorns, and I invite pipeline opponents to saddle up on something that is real."

This was the dynamic McPherson waltzed into when she replaced Linda Duncan in 2019. Thankfully, McPherson had the advantage of being a fresh face without any baggage. On nearly every question but the oil sands, her politics aligned seamlessly with those of her provincial NDP counterparts, who were every bit as horrified by Jason Kenney's Trump-lite approach.

Even on the question of oil production, McPherson struck a far less combative tone than her federal colleagues. She is herself the first progressive child of a decidedly oil-friendly family. Her late grandfather was a well-known oil man in the industry's earliest days; her brother, my old high school pal, for many years owned a company that cleaned rig machinery. And her husband is a lawyer in the oil and gas pipeline industry. He, too, is alarmed by climate change and agrees that the world needs to transition away from oil. "The question for him is the speed at which everything needs to happen," McPherson told me. "That's the point where we vehemently disagree, and for the sake of our marriage we choose not to go down that path."

McPherson was sympathetic to Notley's position on the oil sands, even if she didn't entirely agree with it. "Getting rid of oil and gas today is not going to happen," she said. "We haven't done the work that we needed to do." McPherson sees herself as a bridge, not just between the provincial and federal versions of her party but also between the people of her province and the rest of the country.

"Albertans want a sustainable future," she told me. "The rest of Canada won't believe me, but Albertans are deeply concerned about climate change. But we need to figure out a way to talk about that. Because I'm listening to people talk about 'keep it in the ground,' and all of these very black and white responses to the oil and gas sector and — we can't do that! Those can't be the conversations that we have. It has to be a different conversation and we have to use different words.

"I understand the emergency that we are facing with the climate. But I worry about going too far in that direction and ending up with a Jason Kenney. When you had a Rachel Notley, maybe you didn't like what she was doing, but she had a significant plan that was actually doing more on climate change than any other government in this country, and now we don't. Now we have Jason Kenney. And it's a terrible dilemma to be in."

As we were talking, the U.S. was still months away from choosing the Democratic nominee to confront Donald Trump, and McPherson's dilemma struck me as a powerful encapsulation of the arguments we heard for and against Bernie Sanders, a front-runner at that point.

"This is why we may get Biden, because nobody believes that Bernie Sanders can do it," she said. "As much as we like the idea of the revolutionary who turns things on their head, I don't know if the political system is where we're going to find that revolutionary. It's too careful. Government stopped being radical when social media went up — it's too big a risk."

I said that it seemed to me from afar that Justin Trudeau was caught in this trap. If he were having a beer with us, off the record, I didn't doubt he'd sincerely agree with everything we were saying. But every time he put on his politician's hat, he became tangled up in cautious calculation.

McPherson agreed. She'd interacted with Trudeau a few times in person and had seen him at his best. In January 2020, twenty-seven Edmontonians were among those killed when Iran mistakenly shot down a passenger airliner outside Tehran during the U.S. missile attack; Trudeau flew to Edmonton to meet with their families, many of whom lived in Strathcona. McPherson was there, too, and was impressed by his comportment. "He cried with them, he was super genuine, knew exactly how to listen. It wasn't an act."

But if Trudeau takes naturally to comforting the wounded, he has often seemed incapable of doing the opposite.

"I think of his father, and he was really good at being blunt," McPherson said. "And then I think about would happen if Justin Trudeau tried to do the same. What would be the immediate response if he actually stood up in the House of Commons and said, 'Here's the deal, guys. We are in an emergency and if we don't stop this, we're never going to meet the targets. So this is the plan, we're gonna develop the plan, we're gonna implement it, and it's gonna look like this.' The Conservatives would go *nuts*. They would lose their mind, and they would bring all of the people with them."

"It sounds like you're defending him for not saying those things," I said.

She thought a moment, then replied, "Maybe that's what true leadership is. You say it anyway."

↞∿↠

The next time we spoke in person was six months later, when I took advantage of the pandemic's summer lull to visit Edmonton. It was thirty-two degrees, sweltering for Edmonton and still humid from a week of thunderstorms that had only wandered off that morning to pursue another section of the prairies.

The world had changed radically in the half year since our last encounter. Climate change had fallen off the radar. COVID and Black Lives Matter had seized global headlines, though this deep into Northern Alberta's summer it was possible to feel that nothing much mattered, least of all the WE Charity story then doing the rounds in Canada's newspapers.

I drove to the edge of Mill Creek, where McPherson had just had new solar panels installed on her roof. She showed them off proudly and then we sat in the shade of an enormous willow on her back patio. Her husband and kids were away at a family birthday party; a blind and ancient cat named Eddy (after Sir Edmund Hillary, the explorer) crawled from one of us to the other. A nest full of newborn merlins squawked incessantly from the top of a tall blue spruce.

"Poor mama," said McPherson.

I asked how the past six months had been.

"Crazy," she said. In the early weeks of COVID, McPherson and her staff worked fourteen-hour days, seven days a week. The needs of her constituents were matched by the demands of her federal obligations; because of her roles as the international development and foreign affairs critics, McPherson was part of a team that brought three million Canadians home ahead of the lockdown.

"This one poor girl broke her femur in a motorcycle accident in Thailand — she was nineteen — and then COVID hit. Nobody could get her home. We couldn't get her on an air ambulance because wealthy kids were stealing the air ambulances; their parents were getting them to bring their kids home. We worked really closely with the Liberals, in particular with the foreign affairs minister, who I just kept teasing as Canada's greatest travel agent."

Once that pressure eased, McPherson's attention turned more fully to helping people in Strathcona navigate the complex web of COVID support payments, while simultaneously pressuring the federal Liberals to make those payments more generous and easier to access.

"I was constantly baffled by the fact that we were seeing this *immense* wealth being gained during this pandemic. We're on track to get our first trillionaire with Bezos, and in Edmonton Strathcona people couldn't even afford groceries. They couldn't pay for groceries two weeks in. And the way that the programs were rolled out by the Liberals, it was this little bit here, a little bit there, this patchwork quilt. So nobody knew if they were eligible, nobody knew when help was coming, nobody knew where to turn." Her critique of the Liberal government's blind spots didn't stop her from acknowledging their accomplishments. "It wasn't the most elegant system; however, it was a pretty fast system. Nobody has ever done this, none of us knew what we were doing. So, some credit."

But the country was in chaos, and few people outside the health-care system felt that chaos more keenly than legislators. "Our staff would finish work and then go build little food hampers, because people were phoning our office and saying, 'You know I'm in cancer treatment, I can't leave my house, I don't have anywhere to go.' And we didn't have infrastructure then. Now I think there are more groups doing that and it's a little more formalized, but then there was nothing. There were so many people struggling."

By July 2020, there was a lot of talk about the opportunities this emergency had exposed. Anyone with a radical agenda, wherever they sat on the political spectrum, felt emboldened. "Build back better" had become an international slogan — an easy one to repeat when you're sitting in an armchair thousands of kilometres away from the place laws actually get made. I asked McPherson how it sounded to her.

"This is one of the most exciting times to be a politician, ever," she said. "This is like after the Second World War. What we choose over the next six to twelve months is going to change everything. And there are these huge forces that are going to want us . . . I know this is simplistic, but in my mind, the Liberals just want to get us back to where we were, because that was working okay, so that's where we should go. And the Conservatives clearly feel now's the time: let's shrink government and stop social programming and let's do all those cuts that we've been wanting to do for a long time. And then there's this stimulus, this real investment, where we actually could do those things that we talk about. We *could* actually deal with climate change, we could actually deal with inequality. Even at Foreign Affairs, people are super open to that now; they're recognizing this pandemic as a global issue that requires a global solution. So we *could* make all these great changes."

"But do you think we actually will?"

She sighed. "Depends on the day."

It was impossible to feel on that hot July afternoon, with the pandemic at a low ebb and the world suspended in brief summer ease, but Alberta — still one of the richest societies on Earth — was heading inexorably down its own dark path. The start of the school year was around the corner, and the UCP was refusing to do the bare minimum to help the education system prepare for the inevitable second wave. The very day I met with McPherson, Jason Kenney stood in the legislature and responded to teachers' complaints that they weren't even getting any extra sanitary supplies, let alone janitorial support, by saying, "We expect all the staff in any workplace to help to tidy up, just as we do around here." The medical establishment was getting the same treatment. The next day, the Alberta Medical Association released a poll that found 97 percent of Alberta's doctors had lost confidence in the province's health minister, Tyler Shandro.

The damage was spreading beyond the public sector. Amid devastating cuts to health care and education, Kenney's decision to cut corporate taxes by a third and place a $7.5-billion bet on the doomed Keystone XL pipeline was yielding a parade of disasters. Teck Resources cancelled its $20-billion bitumen mine. Total, the French oil giant, wrote down $9 billion worth of assets and cancelled its membership in the Canadian Association of Petroleum Producers. BlackRock, the world's largest money manager, said it would stop investing in oil sands firms. Deutsche Bank announced it would stop funding new oil sands projects. And on and on.

None of these companies were exiting the oil industry altogether. It was just Alberta they were deserting. That had a lot to do with what Markham Hislop, an energy journalist based in

Calgary, has called the Kenney paradox: By cheering so hard for the oil sands, Kenney became its biggest liability. Kenney and the UCP refused to embrace the kinds of environmental regulations that might have eased investors' consciences, and instead chose to attack anyone and everyone who pushed for such measures (including, in some cases, the oil companies themselves). Kenney's war on Trudeau's carbon tax was the most prominent example but hardly the only one.

Alberta also has a massive problem with tailings ponds and abandoned oil wells, both of which will be leaking contaminants into Alberta's groundwater for decades, if not centuries, unless someone does something about them. Instead of addressing those issues with any kind of policy, or even acknowledging them, Kenney and his spokespeople repeat the "ethical oil" argument, which holds that Alberta's government has a better human rights record than that of other oil-producing countries such as Saudi Arabia or Venezuela. Leaving aside the treatment that First Nations have received in Alberta, the ethical oil pitch entirely avoids the climate question that has so many companies spooked.

In fact, Kenney's entire fiscal policy was predicated on the denial of climate change. This usually went unsaid but not always. Halfway through his term, the UCP's climate denial became explicit thanks to Kenney's high-profile assault on Alberta's environmental NGO community. One of his first acts in office was to launch an official inquiry into the "foreign-funded special interests" Kenney claimed were behind a "political propaganda campaign to defame our energy industry and to landlock our industry." Together with Bill 1, which made it a crime to protest pipelines and other infrastructure projects, Kenney's inquiry put Alberta in the company of those countries he claimed to be more ethical than. A 2019 report from Amnesty International found that over fifty countries have put these kinds of laws in place in recent years. All around the world, at the moment of our greatest ecological peril, the very groups working hardest to preserve the Earth's life systems are coming under systematic attack.

Thankfully, Kenney wasn't yet as good at subverting democracy as Putin or Bolsonaro. The "Inquiry into Un-Albertan Activities," as Markham Hislop dubbed it, dragged on months past its deadline and millions over budget, and produced nothing but unintended irony. One of its exhibits came from a foreign-funded group called Energy In Depth, which is an oil-industry conglomerate devoted to climate change denial. Another equally conjectural report was written by Barry Cooper, a disgraced climate change denier and political scientist at the University of Calgary. In 2008, Cooper was caught funnelling secret donations through research accounts at his university to a non-profit called Friends of Science, which spent the money attacking the Kyoto Protocol and promoting the Conservative Party. In other words, Cooper had done the very thing Kenney was now accusing environmental NGOs of doing, only he'd done it to promote oil production instead of attack it. A third report, by a self-described homeschool teacher living in the U.K. named Dr. Nemeth, set the tone of the entire inquiry by arguing that climate change is no more than a "marketing tool to pursue and achieve a voluntary relatively nonviolent overthrow of capitalism and our current modern industrial society." For this level of insight, the mysterious Dr. Nemeth was paid $27,840 of Albertan taxpayer money.

Not that it was a total waste of time for the UCP. On the strength of its own subtext, the inquiry did contribute to the mass distraction of Albertans from the central problem: The source of their prosperity is now a leading cause of global ruin, a fact to which markets are finally responding. Rather than come up with a plan to deal with this reality, or even acknowledge it, Kenney cut corporate taxes, dropped billions on a doomed pipeline, and discovered "a premeditated, internationally planned and financed operation to put Alberta out of business."

What could McPherson do about any of this? Not much. She could introduce motions in the House, like the one that passed with unanimous consent in August 2020 and committed $2 billion in federal funds to national childcare and back-to-school safety preparations. But she couldn't lead the UCP to reason any more than she could force Justin Trudeau to legislate a Green New Deal. As far as the oil sands and climate change went, her tools were symbolic: public statements delivered via open letter or spoken aloud on Parliament Hill amid three hundred colleagues with their own constituents to worry about. In this way, she could (and did) beseech the federal Liberals to attach conditions to the childcare money they sent to Alberta, so that the UCP couldn't give any of it to private schools; she could (and did) raise a stink about the fact that the UCP had barely touched over $300 million in federal aid meant for essential workers during the pandemic, for fear of having to give credit to the federal Liberals; she could (and did) ask how the hundreds of thousands of Albertans who depend on the oil and gas industry were supposed to transition into sustainable livelihoods without a little help.

Still, real change will only come to Alberta when the United Conservative Party is removed from power or profoundly transformed. McPherson was under no illusions about the likelihood of that happening or her capacity to make it so. But in one of our early conversations, she'd allowed herself to hope that reality would catch up to Jason Kenney and the UCP — that Alberta's electorate would come to see the growing mismatch between their conservative values of hard work, fair play, and fiscal caution and the reckless policies being deployed in the name of those values.

Was it naive to hope that this might in fact be happening? McPherson wasn't sure. "COVID's actually been a bit of a shield" for the UCP, she told me. "We were in free fall in February, and now billions of federal dollars have come into our economy. And those billions of dollars are going to stop at some point, and we are still going to be in free fall. And we have a government that is deeply committed to this path."

Despite that shield, Kenney was the only premier in Canada whose popularity didn't increase during the pandemic. Even if his approach to economic management hadn't caused much backlash among his base, his war against the province's teachers and doctors was a different story. And that was before Aloha-gate: In the opening days of 2021, news emerged that Kenney's chief of staff and six elected UCP members had secretly spent the Christmas holiday vacationing in warm places. This was at the height of the second wave of COVID-19, and these were the very people who'd been warning Albertans not to travel abroad. Among the transgressors was the person in charge of Alberta's vaccine rollout, Minister of Municipal Affairs Tracy Allard, who'd flown to Hawaii with her family. These revelations sparked a truly white-hot public anger that Kenney and the UCP had before never experienced. Kenney apologized for the first time in his premiership; when the storm failed to blow over in three days (people who hadn't seen their own family members in months were not placated by Allard's defence that she'd been loath to break with seventeen years of family tradition), he was forced to demand the resignation of everyone who'd left the country.

Was it a coincidence that Kenney's prospects cratered in lockstep with Trump's final weeks in office? A poll released one week after Aloha-gate showed the UCP's approval rating had collapsed, down 24 percent from the last election, while the provincial NDP's went up 15. Rachel Notley's party had never polled anywhere near the Conservatives; now they were ahead 48 to 31. A week later, Joe Biden cancelled Keystone XL on the day he took office, predictably incinerating the multibillion-dollar bet Kenney had placed using taxpayer money. Kenney responded just as predictably, with outrage instead of repentance, demanding that Trudeau retaliate on his behalf by launching a trade war against the United States. Trudeau, reading a different room, declined.

But all this was still around the corner the last time I called McPherson. It was days before Christmas, and she'd just arrived in Canmore with her family for the first and last week off she expected to have in a long time. It was going to be hard staying offline and quiet, she said, considering the UCP had just sold a few of the nearby mountains to an Australian coal magnate, but she was going to try. "I have a tendency to keep running until I fall down, and this job doesn't allow for falling down." We didn't know what the first week of January held in store; we only knew Joe Biden was about to be president, vaccines were rolling out, and Justin Trudeau would probably trigger a new federal election before long. The political climate was changing, fast.

I reminded McPherson how, almost a year ago, she'd predicted the American public wasn't ready for a radical like Bernie Sanders and that Biden would squeak through instead. I asked her how it made her feel to know she'd been right.

"He's a seventy-eight-year-old white guy!" She laughed. "Even my mom said, 'He's *old*.'" Then she sighed. "I'm delighted he won, because otherwise it was Trump, and I'm so distressed that it was so close. When I think about it too much, it makes me feel a little bit hopeless. The one positive about all this is that I think in fact youth played a very big role. And that's important for us, of course, because if there was an election today and only people under thirty-four could vote, we would have a majority government."

Too bad it was the other way around: whereas youth outnumbers age in the streets, it's age that holds a majority at the polls. As McPherson well knew, the odds of the federal NDP forming a national government were roughly on par with the likelihood of Extinction Rebellion forming a citizens' assembly to run the country's climate policy.

There was something poetic about this, I thought. Maybe it was just romantic. McPherson's solitary political status in Alberta, like her role as a shadow critic in Ottawa, reminded me of those shadow cabinets formed by exiled citizens of dictatorships. These are the people who gather in foreign capitals, preparing for a day when the junta will be banished from their homeland by acting as if that day has come. McPherson was exiled within her own province, which she freely described to me as a "Petro State." She had no legislative power there, and precious little in Ottawa. But that didn't mean she had no influence. Quite the opposite. She had a rare kind of influence, related to that of activists and writers and shadow cabinets, the people who incubate new ideas and push them from the margins to the centre of power. McPherson had more power than those activists, and more freedom than a politician whose party forms the government. In that middle ground, she became a conduit for unconventional ideas. When she advocated for universal basic income or quadrupling Ottawa's international assistance budget, she was normalizing radical proposals, injecting them into the bloodstream of our body politic. Justin Trudeau and the Liberal party will only do something when it ceases to seem radical, and it will only stop being radical when it comes in off the street, puts on a suit, and expresses itself in the House of Commons. In this way, the words of a parliamentarian exert a powerful gravitational pull on the realm of acceptability.

McPherson believed in that power. Despite a year that blew everyone's priorities to smithereens, she still saw one of her greatest strengths as the ability to bridge the conversation between Alberta and the rest of Canada, between environmentalists and the fossil fuel industry. "One of the biggest problems, in my mind, is there doesn't seem to be any space for reasonable discussion," she said. "There's 'shut it all down today' or 'go full steam ahead.' So what will happen is we will go full steam ahead, because that's stasis. That's where we're at right now."

A part of me wants to think that isn't true, that it's simply not accurate to portray the environmental movement as demanding an immediate shutdown of the oil sands. In my mind, we're calling for a halt to expansion, to be followed by an urgent but orderly retreat. We try to stop new pipelines from being built; nobody's tried to shut down the ones already built. But I can see how such nuances slip through the cracks.

In 2018, for instance, a group of land protectors suspended themselves from the Second Narrows Bridge in Vancouver to protest and raise awareness of the Trans Mountain pipeline expansion project and the increased tanker traffic it would bring to the ocean beneath that bridge. One of the protesters was a Greenpeace organizer from Edmonton who later published a story about the experience on Greenpeace's website. "I've seen the damage being done up north in the tar sands," he wrote. "I've seen entire forests turned into barren moonscapes, and lakes full of toxic chemicals — all for the pursuit of oil."

I've written similar things myself. But that kind of language is precisely the kind of conversation killer McPherson was talking about: reducing the complexity of Alberta's prosperity to a tale of greed and destruction. Its primary impact is to give Conservatives like Jason Kenney an opening to present themselves as the only people who see oil and gas workers as decent human beings.

Those workers aren't going to wink out of existence, and neither are Alberta's conservatives. But the elected Conservatives running Alberta are the underdogs now. They've lost the argument, the money's gone, all that's left is anger. That's a dangerous place to be. How can you confront the danger without feeding the anger?

Protest may be necessary, but sheer opposition is no substitute for the vision of an alternative future on which everything depends. Any hope of Alberta shaking free of the grip fossil fuel has on its imagination rests less on the people who suspend themselves from bridges than on those who know how to build new ones.

PORTENTS AND PROPHECIES

CLIMATE ISN'T WEATHER, AS WE always scold deniers when they celebrate a cold day in July. But you can't talk about climate without bringing up the weather, and everyone knows that nothing kills a conversation like mentioning the weather, and there you have the number one obstacle to keeping the weather the way it is: Climate change is boring.

When Nazis or Commies or Jihadis are on the loose, everyone springs into action. Let sea levels rise three millimetres per year, everyone turns the channel. Nobody's trying to kill us in this story, there's no one for us to attack, and that's a problem. The solutions are even more of a buzzkill. Who's going to get excited about geothermal pumps and hydrogen buses? Does it tingle your loins to learn that the United States could cut 70 percent of its emissions just by electrifying absolutely everything — heating, cooking, driving — so long as the electricity was derived from renewable sources? Does your breath quicken when I tell you that we have the technology to do it today, or that doing so would create at least fifteen million jobs, and all we'd have to do is ramp up production of solar panels and wind turbines and Chevy Bolts at the same pace that we ramped up fighter jet production back in World War II? No?

Even so, stories of impending climate doom have twice startled us to the brink of taking the necessary action. Both attempts — first in the mid-1980s, then in 2006 — fizzled out, or rather were extinguished by all of the usual suspects. Now a third moment of international urgency is underway. It started at the end of 2018, when the Intergovernmental Panel on Climate Change (IPCC) published its report on the catastrophic difference between 1.5 and 2 degrees of global warming. That story scared the bejesus out of everyone, in no small part because it said point-blank that people are already dying. Climate change is here and generating quite a bit of startling footage.

Not that we've stopped emitting greenhouse gases, or even slowed. We're dumping nearly forty gigatons of carbon dioxide into the atmosphere each year — a number that has no meaning until placed beside the floods, droughts, hurricanes, forest fires, coral reef die-offs, disappearing glaciers, disintegrating polar ice sheets, and increasing numbers of human refugees and cadavers that have become a staple of our news diet.

This presents a new problem: In the blink of an eye, we've gone from bored to numb.

None of us is anxiously staring out the window at the sky or religiously checking the status of Antarctica's great ice sheet. That indifference, or resignation, or whatever it is, shackles many global leaders who sincerely view climate change as an existential threat to humanity, yet must win the votes of a public whose ignorance has given way to a state of informed delusion as to the magnitude of what's coming.

The most useful thing humans ever discovered is also the most deadly. There's our little paradox, the one we're running out of time to resolve, because the consequences of burning fossil fuels are poised to demolish two centuries of benefit. So how, this time, in our third and final shot, do we make the fight against climate change appealing enough that the people vote for it?

We tell good stories, that's how.

Which brings us back to the beginning. Anyone who tries to tell this tale enters a bizarre hall of mirrors: We've heard it all before and yet need to know more; it's depressing and yet slips into propaganda if we try to inject hope.

Scientists have been slamming into this literary wall for decades. There is no more precise blueprint of our planet's climate, of where it's at and how it's changing, than the Synthesis Report published by the IPCC first in 1990 and updated every five or six years since by thousands of scientists from all over the world. It's all in there and has been for so long that many of the first edition's predictions have become observations. But this bible might as well have been written in Latin. Packed with technical vocabulary and stripped of all emotion, the language of the IPCC reports seems calibrated to deepen our collective slumber, rather than lead the masses from the temptations of oil and gas. And so, as with the Bible, whose imagery is so reflected by the disasters of climate change, you need translators.

←∿→

One thing: Don't look to fiction. There are exceptions, absolutely, but for the most part — and this is truly shocking — that class of artists which most prides itself on serving as a sentinel of the human experiment has missed this story completely. "Fiction that deals with climate change is almost by definition not of the kind that is taken seriously by literary journals: the mere mention of the subject is often enough to relegate a novel or a short story to the genre of science fiction," writes Amitav Ghosh in *The Great Derangement*, his brilliant (non-fiction) inquiry into fiction's

Earth-sized blind spot. "It is as though in the literary imagination climate change were somehow akin to extraterrestrials or interplanetary travel." Ghosh includes himself in this damning assessment, being the author of several acclaimed novels that also failed to mention climate change. (He has since rectified this oversight with his 2019 novel, *Gun Island*.)

"In a substantially altered world," he predicts, "when sea-level rise has swallowed the Sundarbans and made cities like Kolkata, New York, and Bangkok uninhabitable, when readers and museumgoers turn to the art and literature of our time, will they not look, first and most urgently, for traces and portents of the altered world of their inheritance? And when they fail to find them, what should they — what can they — do other than to conclude ours was a time when most forms of art and literature were drawn into the modes of concealment that prevented people from recognizing the realities of their plight? Quite possibly, then, this era, which so congratulates itself on its self-awareness, will come to be known as the time of the Great Derangement."

One of the problems Ghosh identifies is that many of climate change's most harrowing manifestations are simply too implausible for novelists to get away with putting in a story. When Ghosh was a young man, he was walking through Delhi to visit a friend when a freak tornado touched down on the street he was on. He dove behind an apartment balcony for shelter, and the eye of the tornado passed directly over him, wrapping him in a moment of otherworldly calm. When it passed, he emerged into a scene reduced to rubble; thirty people were killed and seven hundred more were wounded.

"No less than any other writer have I dug into my own past while writing fiction," Ghosh says. "By rights then, my encounter with the tornado should have been a motherlode, a gift to be mined to the last little nugget." And yet try as he might, and he did for decades, Ghosh was never able to recreate that scene or any version of it in his fiction. "In reflecting on this, I find myself asking, what would I make of such a scene were I to come across it in a novel written by someone else? I suspect that my response would be one of incredulity . . . Surely only a writer whose imaginative resources were utterly depleted would fall back on a situation of such extreme improbability?"

When I look back on western North America in September 2020, there were any number of such scenes. Smoke shrouded the city I live in, Vancouver, for over one straight week. The streets of San Francisco, Portland, and Seattle were plunged into murky orange like a set from *Blade Runner*. Elsewhere in the world, the scenes of climate change have been equally dramatic — and therefore equally challenging to capture in a novel. "The calculus of probability that is deployed within the imaginary world of a novel is not the same as that which obtains outside it," Ghosh writes. "This is why it is commonly said, 'If this were in a novel, no one would believe it.' Within the pages of a novel an event that is only slightly improbable in real life — say, an unexpected encounter with a long-lost childhood friend — may seem wildly unlikely: the writer will have to work hard to make it appear persuasive.

"If that is true of a small fluke of chance, consider how much harder a writer would have to work to set up a scene that is wildly improbable even in real life? For example, a scene in which a character is walking down a road at the precise moment when it is hit by an unheard-of weather phenomenon?"

That's merely the beginning of a fascinating investigation of why fiction writers have, for the most part unconsciously, avoided climate change. If you were wondering about that, *The Great Derangement* is the book for you.

But if you want to read about climate change, best stick to journalism.

Of course, sometimes novelists also make the best journalists. Among them is Nathaniel Rich, whose non-fiction book *Losing Earth* offers a character-driven examination of why it took us so long to take climate change seriously. As Rich makes painfully clear, we have been here before: "Nearly every conversation that we have in 2019 about climate change was being held in 1979." It's heartbreaking to learn how close the United States once came to leading this global fight. Like an addict looking back on the way things could have been, *Losing Earth* zeroes in with cinematic detail on the efforts of two Americans to raise climate change from fringe issue to presidential priority over the decade ending in 1989.

One of those Americans is now the most venerated climatologist on Earth, former director of NASA's Goddard Institute James Hansen; when we meet him in 1979, he is a brilliant but media-shy scientist who "was not afraid to follow his research to its policy implications." The other is Rafe Pomerance, a charismatic and Washington-savvy environmental lobbyist searching for the right messenger to tell the world about climate change. He finds what he's looking for in Hansen, whose early reticence gives way to an urgent clarity verging on prophecy.

Rich follows the two men as they pierce the opaque rings of power and connection encircling Ronald Reagan's White House. It's startling to learn that as early as 1983, "the issue [of taking action against climate change] was unimpeachable, like support for the military and freedom of speech." But Reagan wasn't having it. Pomerance and Hansen had to wait for Reagan's second term to wind down before they got their breakthrough. In 1988, Hansen, by then a trusted household name, delivered spellbinding congressional testimony on what happened to be the hottest June 23 in Washington, D.C.'s history, exploding the issue of climate change onto an American public that was still congratulating itself for tackling the "hole" in the ozone "layer." (There was never any hole, Rich explains, and no layer either; ozone molecules are evenly distributed throughout the atmosphere, but the lethal drop in their concentration above Antarctica appeared as a void in satellite imagery. After one of the chemists who discovered this casually described it as an "ozone hole" during a slide show, the press found its catchphrase and never let go. It was among the most successful — albeit accidental — environmental marketing campaigns in history.) Hansen's testimony spurred George H.W. Bush to campaign on a promise to fight the greenhouse effect with "the White House effect." But the fossil fuel lobby outmanoeuvred their environmental counterparts in Bush's subsequent administration, and the true White House effect was to quash aggressive climate policy both at home and abroad in the name of economic pragmatism.

Losing Earth is a gorgeously crafted story, revelling in dialogue and human foible, loaded with foreshadowing of today's climate change debate. When it first appeared in August 2018 as an extended article that filled an entire issue of *New York Times Magazine*, *Losing Earth* went viral. But many environmentalists took issue with Rich's conclusion that our failure to reduce emissions has more to do with human nature than Republicans or oil companies. Rich took some pains to address those criticisms in the book-length version, emphasizing that governments that ignore climate change are guilty of "crimes against humanity." But he still insists that we've been willing dupes. Climate change, he points out, has been graphically explained in major print and television media since at least 1953, when *Time*, the *New York Times*, and *Popular Mechanics* all ran articles on the subject. Every major American utility and auto company studied the problem extensively in the 1970s; even the environmental community waited until the late 1980s to press an issue it had known about for over a decade. "Everyone knew," Rich writes, "and we all still know."

His point isn't that we're doomed by our intrinsic bias for short-term reward over long-term consequence. It's that knowledge alone is never enough. The mistake Hansen and Pomerance made was to assume information alone would galvanize the public. But the challenge, then as now, isn't merely to inform; it's to awaken. "When popular movements have managed to transform public opinion in a brief amount of time, forcing the passage of major legislation," Rich writes, "they have done so on the strength of a moral claim that persuades enough voters to see the issue in human, rather than political, terms." If you want to talk about climate change, he concludes, "The first requirement is to speak about the problem honestly: as a struggle for survival."

<center>~~~></center>

The second time humans almost beat climate change — or, at least, were united in an effort to try — was brought to us by *An Inconvenient Truth*. On one level, Al Gore's 2006 opus could be seen as a counterargument to Rich's point that knowledge alone is not enough: Here was a movie that told us what we already knew, and it worked. For a year, climate change shot to the top of the news. Riding the renewed wave of public concern, Gore founded the Alliance for Climate Protection, which gathered so much bipartisan support that none other than Newt Gingrich sat before TV cameras on a couch with Nancy Pelosi to declare, "Our country must take action to address climate change."

Of course, that moment sputtered, too, much more quickly than the last one and without ever reaching the same heights. But the lesson it offered was real: A good way to wake someone up is to slap them in the face. Yes, we knew all about climate change in 2006, but nobody had compiled such a cinematic collection of apocalyptic horrors and shoved them under our noses quite like that before. It helped that the person who did so had once won the popular vote for the American presidency. An even greater boost might have come from Hurricane Katrina, which just the year before had primed the world's imagination with graphic and unprecedented footage of a great American city laid to waste. When it comes to climate change, seeing is believing.

That seems to have been the literary lesson that David Wallace-Wells took to heart when writing *The Uninhabitable Earth*, which reads like a 228-page elaboration of his opening sentence: "It is worse, much worse, than you think." One of Wallace-Wells's central insights is that we're confusing a best-case scenario with its opposite. "As recently as the 1997 signing of the landmark Kyoto Protocol," he writes, "two degrees Celsius of global warming was considered the threshold of catastrophe." Twenty years of unrestrained fossil fuel consumption later, "two degrees looks more like a best-case outcome, at present hard to credit, with an entire bell curve of more horrific possibilities extending beyond it and yet shrouded, delicately, from public view."

The Uninhabitable Earth rips that shroud away. Taking two-degree warming as a starting point, Wallace-Wells assails us with a minutely detailed, relentlessly escalating catalogue of consequence. Here's a typical passage:

> At two degrees, the ice sheets will begin their collapse, 400 million more people will suffer from water scarcity, major cities in the equatorial band of the planet will become unlivable, and even in the northern latitudes heat waves will kill thousands each summer. There would be thirty-two times as many extreme heat waves in India, and each would last five times as long, exposing ninety-three times more people ... At three degrees, southern Europe would be in permanent drought, and the average drought in Central America would last nineteen months longer and in the Caribbean twenty-one months longer.

If only that were the bad news. According to the IPCC, as of 2018 our emissions trajectory put us on a path for four degrees by the end of this century. "That would deliver what today seems like unthinkable impacts," Wallace-Wells writes. "Wildfires burning sixteen times as much land in the American West, hundreds of drowned cities. Cities currently home to millions, across India and the Middle East, would become so hot that stepping outside in summer would be a lethal risk."

Judging by its success, that kind of straight talk still has the power to captivate an audience. When *The Uninhabitable Earth* first appeared in 2017 as a magazine article, it quickly became the most widely read story in *New York Magazine*'s fifty-year history. Just like *Losing Earth*, Wallace-Wells's framing of the problem provoked criticism from some environmentalists, who accused him of sensationalizing the issue by cherry-picking the darkest possibilities. The environmental magazine *Grist* responded with the emblematic headline, "Stop scaring people about climate change. It doesn't work." Overstating the probability of worst-case climate outcomes, the thinking goes, could feed deniers the kind of exaggerations and erroneous predictions they love to wave as evidence that climate science is bogus. Worse, a relentless focus on horrific outcomes could paralyze readers with a sense of helplessness.

But *not* scaring people about climate change doesn't work, either. Especially in the West, which hasn't seen total calamity since World War II, the prospect of civilizational collapse has come to seem impossible. "We suffer," Wallace-Wells writes, "from an incredible failure of imagination."

That's no slight on the IPCC's contributing authors. It was their 2018 report, after all, that kicked off our present moment and continues to be interpreted and amplified by the usual globe-spanning army of translators. But they all got a signal boost from the same coincidence that helped ignite *An Inconvenient Truth*.

The summer of 2018 happened to be the most destructive wildfire season in California's history (at the time — it was dwarfed in 2020), and the ashes were still smouldering when the IPCC report dropped on October 8. One month later, while writers were still busy disseminating its findings, the Camp fire erupted in northern California like an exclamation mark. That fire destroyed a town called Paradise, killing eighty-five people and causing as much as $16.5 billion in damages. Nature is starting to do a lot of the translating for us. But if a disaster can illuminate the problem, it takes human imagination to articulate solutions.

Enter Naomi Klein, climate evangelist. Her 2019 book, *On Fire: The (Burning) Case for a Green New Deal*, evokes the most hopeful aspect of our third attempt: the likelihood that America's forty-sixth president will finally follow through with George H.W. Bush's abandoned promise from 1988, and then some. "Our current moment is markedly different" from the previous two climate interventions, Klein writes, for two reasons: "one part having to do with a mounting sense of peril, the other with a new and unfamiliar sense of promise."

Klein tracks both halves of the equation through a collection of new and previously published articles that track a series of climate disasters and the responses they inspire: Deepwater Horizon; a climate-change-deniers conference; the Pope's radical call for environmental preservation; the hurricane that laid waste to Puerto Rico in 2017; and the Green New Deal, which went on to be endorsed by every Democratic candidate for president except the one who won. Arguably, though, even Biden's climate plan is a Green New Deal in all but polarizing name.

Most of Klein's new material is found in the book's essay-length introduction and epilogue, which introduce rising characters on the brink of becoming climate leaders. The former is where we meet Greta Thunberg, who, naturally, is now a friend of Naomi Klein's. However skeptical you may be of a teenager's capacity to grasp the geopolitical complexities of decarbonization, it's incredible how far Thunberg has gone since the day in August 2018 when she first skipped class to stand alone outside of Sweden's parliament with a handmade sign. Klein's epilogue turns to another prodigy who had recently exploded onto the world stage: Alexandria Ocasio-Cortez, co-sponsor of the Green New Deal and many more things besides.

Depending on where you stand, Klein can either come off as the whole-earth theorist we need in a time of global collapse or a dangerous purveyor of utopia. She's well aware of the mixed reactions she inspires. One of her more revealing essays, "The Leap Years," ruminates on her self-described attempt to insert the Leap Manifesto — "a kind of proto–Green New Deal" — into Canadian politics back in 2015. To summarize the quixotic attempt: Klein and her co-collaborators got laughed out the door.

But here's the thing about Klein and the Green New Deal both: If the far left's call for a complete overhaul of modern capitalism carries a whiff of ludicrous overreach, how should we characterize society's response? Consider for a moment the global slaughter of our Anthropocene era — 85 percent of land mammals wiped out, another million species on the brink of extinction, humanity itself now at risk — and ask yourself who is more delusional: Naomi Klein or the pragmatists minding the status quo?

"Ours is an age," Klein writes, "when it is impossible to pry one crisis apart from all the others." Anyone who's paying attention must agree. This story is much older, and much bigger, than humanity's love affair with fossil fuels. More explicitly than any other writer, Klein talks about climate change as an organizing principle for the all-encompassing crises of our age. This is both a genuine insight and a clever way to sustain attention in a media environment saturated with emergencies. Her fear isn't that we'll fail to tackle climate change in what little time remains (though there's that); it's that we'll replace one system of oppression with another, trade the world's internal combustion engines for a billion Teslas, only to carry on shredding the biosphere.

Any critique of Klein's analysis ought to begin with an acknowledgement of that central insight and maybe some appreciation of the risk involved in daring to propose solutions. After that, by all means, explain why solving climate change is a big enough challenge on its own without adding racism, inequality, and plummeting biodiversity to our to-do list. Point out that California used modern capitalism to lower emissions below 1990 levels by 2016 — four years ahead of schedule, and they're still dropping. Feel free to believe that the just, sustainable society for which Klein has advocated all her life is neither realistic nor remotely literary. Utopia, it's true, fails the suspension-of-disbelief test that every good story must pass.

Just keep in mind that this is a time when real life is providing plot twists that no self-regarding editor would allow past a first draft. Many of those twists are dark and Trumpian, like a burned-out town named Paradise. But there's no objective reason they couldn't go the other way; no reason, for example, that something as outlandish as a Leap Manifesto couldn't hop across the border and become something that looks an awful lot like a Green New Deal.

One thing's for sure: whatever happens next will be dramatic. If only we could read today what climate journalists will be reporting in a decade. Then again, we'd probably write it off as science fiction.

A BRIEF HISTORY OF POPULATION CONTROL

THE OTHER DAY I SAW the following pronouncement on a billboard in Vancouver: "The most loving gift you can give your first child is not to have another." Had the message been solely in text, I might have been tempted to agree. My wife and I chose to stop at one, though that was more about our own carrying capacity than the planet's. Still, there *are* quite a few people on Earth, quite possibly too many. Alas, the billboard's message wasn't limited to text. It included a very large photo of a cute little baby who happened to be Black.

The advertisement was part of a campaign by a U.S.-based non-profit called One Planet, One Child, whose mission is to "alert and educate that overpopulation is a root cause of resource depletion, species extinction, poverty, and climate change." After the very predictable uproar that greeted its Black-baby-reducing billboard, the group immediately took it down and issued one of those non-apologies that have become a staple in our culture. You know, the kind that regrets the misunderstanding without demonstrating any remorse.

One Planet, One Child insisted this particular misunderstanding arose as a result of how sensitive to racial issues they really are. "We endeavored to be as inclusive as possible in this ad campaign," the group said, "so we were running six variations of this ad, depicting children or families of different nationality or skin colour. We want to include everyone in the conversation." Their only mistake, they said, was "that we didn't anticipate the possibility someone might see only the one ad, or that someone might share just that ad on social media."

Given that the billboard with the Black baby wasn't put up anywhere near any of the others, the real question is how they failed to anticipate that the public would see what we were being shown. Which was, to be clear, the suggestion that some people have a greater right to exist than others.

The ordeal did at least serve to illustrate why most environmentalists steer clear of the subject of overpopulation. We might discuss it in general terms, but we're loath to offer specific prescriptions. It's not that overpopulation isn't a planetary concern. It's that the moment you say humans need to make fewer babies, you run into the question of who exactly should be making fewer babies, and who should be telling them to do so. And because suggestion alone tends not to have much influence over people's baby-making habits, your arguments are bound to wander into coercion territory, which is never a good look. By now your rhetoric is merging with a long immoral history of forced sterilization programs and racial discrimination whose slippery slopes invariably funnel straight down to conclusions that all bear a troubling resemblance to the Final Solution.

↞∿↠

Ever since Thomas Malthus published his *Essay on the Principle of Population* in 1798, the elite's fears of the unwashed masses and non-whites overrunning polite society have been dressed up as concern for the environment. When Malthus wrote "the power of population is indefinitely greater than the power in the earth to produce subsistence for men," he was less concerned about the British countryside's carrying capacity than he was about the fact that farm workers were reproducing faster than aristocrats. Rather than allow poor people to start piling up, Malthus — who found the idea of contraception and sterilization repugnant — advocated a different means of keeping their numbers down:

> In our towns we should make the streets narrower, crowd more people into the houses and court the return of the plague. In the country, we should build our villages near stagnant pools and particularly encourage settlements in all marshy and unwholesome situations. But above all, we should reprobate specific remedies for ravaging diseases, and those benevolent but much mistaken men, who have thought they were doing a service to mankind by projecting schemes for the total extirpation of particular disorders.

It was a short leap from Malthusian principles to social Darwinism and eugenics. *On the Origin of Species by Means of Natural Selection* was published in 1859, with a subtitle we tend to forget: *The Preservation of Favoured Races in the Struggle for Life.* Darwin himself didn't spend much time producing or consuming racist literature — his interest was in non-human species — but he remained a product of his culture. His 1871 tract, *The Descent of Man*, did include the casual prediction that "at some point, not very distant as measured by centuries, the civilized races of man will almost certainly exterminate, and replace, the savage races throughout the world."

Whereas Darwin was content to let nature run its course, his wealthy cousin Sir Francis Galton argued well-heeled whites should do their best to hasten the process along. It was Galton, a fellow scientist and much inspired by Darwin, who coined the term "eugenics," from the Greek "eugenes," meaning roughly "of good stock." "We greatly want a brief word to express the science of improving stock," Galton wrote, describing his vision of selective breeding that would "give the more suitable races or strains of blood a better chance of prevailing speedily over the less suitable than they otherwise would have had." At least Galton was less murderous than Malthus. Rather than inflict plague on the lower classes and withhold medical care, he encouraged "suitable" people to have more children and outbreed their "unsuitable" counterparts. But as time went on and the "overgrowth of population" continued unabated, he began calling for mass sterilization of the worst of the unsuitables.

That set the tone for the next half century as the movement Galton founded spread throughout the white world. Eugenics crossed the ocean to find a fertile garden in North America's upper classes. Its most influential proponent was arguably Theodore Roosevelt, who used the power of the presidency to promote sterilizing criminals and the "feeble-minded," and who publicly mused that the United States would be committing "race suicide" if eugenics weren't taken up more aggressively. Back in England, Winston Churchill was soon admiring the fact that "60,000 imbeciles, epileptics and feeble-minded" were sterilized against their will in the United States between the two world wars. In the end, it took the Nazis to give eugenics a bad name. In their pursuit of lebensraum for the "master race," the original eco-fascists sterilized half a million people before deciding it wasn't just the unborn who should be denied life but their living seed-carriers, too.

The Holocaust put an end to that kind of talk but not to the sentiment. Sterilization of minority groups and marginalized people remained an international practice long after the war ended. In Canada, Indigenous women continued to be sterilized against their will, and sometimes without their knowledge, through the 1970s, a decade in which some 1,200 Indigenous women suffered the procedure. It was worst in the north: in Inuit communities (in what is now Nunavut), up to a quarter of adult women were sterilized in government hospitals. That ended after the practice sparked a national outcry and subsequent government inquiry, but Indigenous women have continued to be surreptitiously and haphazardly brutalized in this way across Canada ever since. In 2019, over one hundred Indigenous women from six provinces and the Northwest Territories signed onto a class action lawsuit alleging they'd been sterilized without informed consent between 1985 and 2018.

For all their heartless zeal, the eugenicists failed miserably. The twentieth century saw humanity's growth chart go vertical. We just exploded across the planet. The human population grew from 1.6 billion in 1900 to over six billion by 2000, an order of magnitude more growth than the previous century had seen, with the growth rate peaking in the early 1960s.

In 1964, Paul Ehrlich's environmental blockbuster *The Population Bomb* was published, opening with the statement: "The battle to feed all of humanity is over. In the 1970s hundreds of millions of people will starve to death in spite of any crash programs embarked upon now." Ehrlich essentially reprised Malthus's argument without being quite so overtly classist. He feared that the human population was on a collision course with the planet's carrying capacity. *The Population Bomb* was a plea to save ecosystems as much as people — it was commissioned by the Sierra Club — and Ehrlich did have the good grace to say that the United States should lead the way in reducing its own population, both because Americans had the highest environmental footprint and also because doing so would give Americans the moral authority to play a global leadership role on this issue. It was in his vision of global leadership that things got sketchy. Ehrlich advocated a sort of international triage whereby wealthy nations, such as the U.S., would provide food aid only to countries with a solid plan for reducing their populations to the point of self-sufficiency; countries that didn't have any such plan would be left to their own devices so that famine could control the population. Better they starve now, Ehrlich argued, while the population was still relatively low, than later when the population had doubled and the inevitable collapse in food supply afflicted twice as many people.

One of the countries Ehrlich thought beyond redemption was India. His visit to that country featured prominently in *The Population Bomb*. He described the horror of driving through Delhi: "People thrusting their hands through the taxi, begging. People defecating . . . people, people, people." Ehrlich thought he was describing the effects of overpopulation, but Delhi was no more densely populated than New York City. What he was looking at was poverty. "I don't see how India could possibly feed two hundred million more people by 1980," he wrote.

That didn't stop India from doing so, and then some. Even as Ehrlich was writing, the Green Revolution — also known as the third agricultural revolution — was underway. It doubled crop yields through a combination of new synthetic fertilizers, pesticides, high-yield varietals of cereal crops, and intensive irrigation. Fifty years after Ehrlich's prediction of mass starvation, it still hasn't come to pass.

Still, it has to be said that Ehrlich's concerns were not imaginary. The industrialization of Indian agriculture that fuelled a near quadrupling of the population, from 360 million in 1947 to 1.4 billion today, has had its own unintended consequences. Among the worst is that the country's vast monsoon-charged aquifers have been sucked dry. Those modified crops whose incredible yields Ehrlich failed to foresee are also incredibly thirsty; a quarter of all the world's groundwater extraction now occurs in India, and lethal water shortages have placed farmers in many parts of India in a state of perpetual crisis (exacerbated, to be sure, by the predatory grip of companies like Monsanto on their seed supply). The monsoon does replenish India's aquifers each year, but it's clear that more water is being drawn out of the system than the rains can return. And those rains are increasingly unpredictable as climate change wreaks havoc on a weather pattern that formed the basis of India's great civilization.

As Ehrlich and many others have argued, the population explosion of *any* species inevitably impacts the landscapes in which it occurs. China isn't down to its last fifty tigers because of European driving habits. In light of its own environmental deterioration and spurred by an ingrained cultural memory of the devastating famine that Mao presided over, China was one of the countries that took the message of *The Population Bomb* to heart, implementing a two-child policy in 1969 that was tightened ten years later to one child per family. That program succeeded in stopping China's population growth. It also triggered an explosion of sex-selective abortions, leading to an estimated thirty million more marriageable men than women by 2020. And it further entrenched a political philosophy of coercion that includes the forced sterilization of China's Uighur Muslims as part of an ongoing campaign of cultural genocide.

One thing China's population control didn't do is save the environment. As of 2019, China led the world in carbon emissions, emitting twice as much as the runner-up (United States) and four times as much as India, despite having almost the exact same population. The deciding factor in China's rising emissions wasn't population but industrialization.

The problem isn't how many people live in a given place; it's how they live. Even with the Chinese population's meteoric improvement in living standards, their overall environmental footprint (accounting for consumption of all resources, not just fossil fuel) remains tiny compared to North Americans'. According to the Global Footprint Network, we use up the equivalent of 8 hectares per year per person. In China, it's 3.7 hectares; in India, the figure is 1.2; for most African nations, it's well under 1.

The question arises: Don't we middle class citizens of the industrialized world want everyone on Earth to enjoy our living standards? Isn't that the very definition of equality, the goal to which all our best instincts reach, implied in every campaign to eradicate poverty? But that would be disastrous for the planet. The planet's forests, oceans, freshwater sources, and atmosphere would all be torn asunder if four more billion people started living the way we do. Actually, those things are already being torn asunder. Not because of the world's poor, but because of us.

It all gets messy, fast. I'm not sure what the answers are. All I know is which way the numbers and my moral compass point: Our insatiable "Western" appetites for consumption are more urgent problems than overpopulation. If you live in an industrialized country and environmental impact is truly your concern, it's your own lifestyle you should be thinking about, not the reproductive habits of foreigners.

←∿→

Of course, the marketing team at One Planet, One Child would say that was precisely the point they were trying to make — their billboard was in Vancouver, after all, not downtown Delhi. It was *our* reproductive habits they wanted us to think about, not anyone else's.

But North Americans have already absorbed that message. As in just about every country in the developed world, our birth rate has dropped below the level a population requires to maintain itself. Immigration is the only reason our population isn't shrinking. Declining birth rates are in fact a global trend; almost everywhere on Earth, people are having fewer children, so that for the first time in human history, our species' rate of population growth is slowing down. And we're doing so of our own volition. No plagues or famines, wars or mass sterilization programs required. No billboards, either.

This is a wonderful story, one of the best stories the environmental movement can and should be telling, because it's based on a magical ingredient that merges human and ecological well-being: the empowerment of women. It turns out that when girls are educated alongside boys and women are not blocked from full participation in society, plus given access to contraceptives, they tend to have fewer children. They also start families later in life. It's a golden, global rule.

This is win-win fairy-tale material: As a result of women's empowerment, the world's population is settling down. Tell everyone you know: Donating to a women's health NGO is an environmental act.

To be clear, the global population of humans is still rising in absolute terms and will for some decades to come. But our reproductive rate peaked a long time ago, in 1961 at 2.1 percent, to be precise. At that point, our numbers were doubling every thirty-three years. Cut Ehrlich a little slack for being worried. Today, our growth rate is just barely over 1 percent, which would see us doubling in seventy years if we kept on at this rate, but we won't. Our growth rate has been slowing for sixty years and will keep doing so until the global population plateaus and finally starts to contract. Demographers the world over agree on that trajectory, and disagree only on the details.

What will our peak population be, and when will we reach it? Many institutions, including the United Nations Development Programme, have estimated we'll peak at around ten billion by 2100. But models are updated every year, and they keep revising their peak estimates downward. In 2020, *The Lancet* published a major study that didn't get the press it deserved because, well,

2020. The authors predicted that global population would peak in the year 2064 at 9.73 billion, and then start declining. Most countries on Earth (151 of them according to this study) will peak well before then, by 2050. That remains an educated guess, but it bears repeating that the common factor linking this *Lancet* story to every other study like it is women's education and access to contraception.

My great-grandparents, probably like yours, all had between five and ten children. Many of those kids died before reaching adulthood or soon after. That's how it was for pretty much all of human history — the reason it took Homo sapiens fifty thousand years or so to make their first billion is that, until the twentieth century, child mortality and disease and war and famine killed off almost as many of us as we could replace.

Each of the generations that followed my great-grandparents had half as many children as the one before. Now I'm the father of a single child who happens to be a girl. I'm not trying to tell any-one else how many kids they should or shouldn't have; whatever you decide, it won't affect the curve eight billion of us are all on. All I'm saying is I'm glad my daughter will be the one who decides her own fate. If the fate of the world rests on women everywhere being able to do the same, that's a vision of population control I can get behind.

DANGEROUS OPPORTUNITIES

January 20, 2021

THE PAST SEVENTY-EIGHT DAYS HAVE resurrected, yet again, the immortal words of Antonio Gramsci, founding member and one-time leader of Italy's Communist Party: "The crisis consists precisely in the fact that the old is dying and the new cannot be born; in this interregnum, a great variety of morbid symptoms appear."

I first read that long before Trump slunk out of the White House this morning, and it rang true even then. I suppose people have always felt they were living on the verge of a new era. They've probably always been right. If you really want to be philosophical about it, you could say life itself is an interregnum, and the world remakes itself every single day.

But that's not what Gramsci had in mind. He wrote those words almost a century ago, in prison, having run afoul of Mussolini and the fascists. They threw him in jail from 1926 to 1935, during which time he wrote three thousand pages of historical and political analysis in secret. He had to smuggle the notebooks in and out of his cell. The *Prison Diaries* weren't published until the 1950s, long after Gramsci's death in 1937. He never lived to see the end of his own interregnum. Over fifty million people were slaughtered before the old was finished dying and Europe's new peace was born.

These kinds of stories help me keep my own life in perspective, though you can't go around being constantly grateful you're not rotting in some fascist dungeon. Sometimes it's okay to let yourself feel overwhelmed by the morbid things happening around you, even when they're not (yet) happening directly to you. It's okay to acknowledge all the slow dying, and painful birthing and irreversible suffering that happens in between, imposed on all the people and animals and plants for whom posterity will be about as much consolation as the Marshall Plan was to victims of the Holocaust.

I've never known such horrors, but for all my life they've seemed to hover just past the horizon. The Cold War, during childhood, with its corollary of nuclear winter. The Rwandan genocide, the summer I graduated high school. The prospect of ecological calamity had risen to global prominence by the time 9/11 struck; just when the world might have turned its attention to climate change, the winner of the Cold War chose to launch an oil war in the Middle East. Those Bush years were one long morbid symptom. Obama's presidency did at first resemble a cure; the election of a Black president felt like the birth of a new era, but almost right away the baby got bogged down in the muck of institutional inertia and Republican bad faith. Wall Street and the military-industrial complex kept on making killings. Pretty soon, the birther fiction was being peddled by you know who, and we were officially back in interregnum land.

Meanwhile, we Canadians lost an entire decade to Conservative rule under Stephen Harper. I admit it seems a little quaint today to think of what used to pass for bleak, but at the time our creeping Petro State and its proud contempt for environmental concerns and Indigenous rights were enough to knock the air from our progressive lungs. We had a brief emotional respite after the Liberals' victory in 2015 (remember the Paris Agreement, when team Trudeau and Obama made it seem as though the world would take on climate change?), but then America's white-lash forced Trump upon the world. In times like that, Canada always feels more like an American protectorate than a British colony. For four teeth-grinding years, we could only hope that the Trump administration was just another one of Gramsci's symptoms: morbid, but fleeting.

Then came the election. Then January 6. The old was dying but wouldn't let the new be born, and boy did a great variety of symptoms ever crawl out of the woodwork. In the seventy-eight days between November 3, 2020, and January 20, 2021, Gramsci's metaphor crystalized into a hard diamond, the most flawless interregnum I've ever seen. By the time you read these words, the world will have a better understanding of that diamond's worth — how much it cost, I mean — but one thing that might fade with hindsight and analysis is the way it feels right now.

It feels hopeful. Not just because the new administration has finally been born but because of how effectively the previous one just killed itself.

It's horrifying, too, of course. But it's been horrifying all along, without any of the relief these final days have provided. Because, strange as it sounds, it really was a relief to see all those thousands of symptoms flush themselves into the open on January 6 and attack the Capitol, convinced they were the cure. Five people died, many were maimed, but it could have been so much worse. It could have succeeded. Instead, it was a debacle. A disgrace so total it exposed everything that had been going on for the past four years, and the two hundred before that, in a way that no amount of journalism or fact-checking ever could.

Utter mendacity, white supremacy, conspiratorial delusion, bloodlust, sedition. Those descriptions, which only begin to convey the thing calling itself Trumpism, are so shocking to middle-class sensibilities that until we saw the mob, literally howling for blood, with our own eyes it was still possible for an absurdly large proportion of the public to ignore the barbarity at the heart of the Republican party, half-heartedly concealed behind the leadership's tailored suits and political catchphrases. Now it's bull horns and fur costumes, Confederate flags and Camp Auschwitz shirts, a lusty chant to "hang Mike Pence." This is the road they've been hauling us down all along. This is where it leads. They will never admit it, but they can no longer deny it.

It's not just an American thing. Canadian Conservatives are caught out, too. They'd been flirting with the far right for years, just less brazenly. "Canadians relieved our right-wing, authoritarian politicians not charismatic enough to inspire coups," ran *The Beaverton's* headline a few days after the Capitol attack, above a photo of five Conservative leaders swiped from an infamous cover of *Maclean's* magazine. In the lead-up to Canada's 2019 election, the Conservative Party channelled Trump to warn Canadians on its website that "Justin Trudeau is rigging the next election." Andrew Scheer, Trudeau's Conservative opponent in that election, hired as his campaign manager the former director of Rebel Media, the closest thing we have to Fox News. After Conservatives lost that election, Erin O'Toole ran his winning campaign to replace Scheer as leader of the federal Conservatives on the Trumpian slogan "Take Back Canada." O'Toole also hired his own unsettling campaign manager, the founder of an organization called Ontario Proud, which is just as far right as it sounds. (The Proud Boys were founded by an Ontarian, Gavin McInnes.) The Conservative Party's deputy leader, Candice Bergen, has been photographed in a camo MAGA hat. Jason Kenney went to D.C. just before the pandemic hit and took delighted selfies with fellow Calgarian Ted Cruz.

Not until after the coup failed so embarrassingly did Canada's Conservatives start swimming the other way, with O'Toole clarifying that there's "no place for the far right" in his party. One hopes they have indeed been scared a little straight, but O'Toole's still a long way from acknowledging the obvious truth, which is that there has in fact been a place for the far right in Conservative quarters. In a multiparty system where a third of the vote is enough to form the government, Conservatives have long seen a path to victory in doing what no other party in Canada will: ignore climate change, pooh-pooh abortion and Indigenous rights, incite separatists in Alberta, and aggressively smear their opponents as woke, elitist pansies.

However! The last two and a half months really sucked the oxygen out of that rhetoric. By resorting to mob violence on January 6, the Trumpkins effectively admitted they'd lost the argument. They ran out of words and started punching. The obvious question now is whether the violence has only just begun. But mixed in with that foreboding concern — which, remember, isn't new — is a significant amount of hope. That *is* new. It is born of a widespread acknowledgement of reality.

The gloves are off, the jig is up, the witch is dead. Biden and Harris and the most diverse, experienced cabinet in American history are taking the reins, determined to vaccinate their country, decarbonize their economy, tackle inequality, and pull racism out at its roots. They have Congress and the Senate behind them, and the army, too. As for Canada, progressives are newly emboldened while Conservatives are reduced to spluttering, "That's not us." If an election were held tomorrow, it wouldn't go well for them.

Good news for progressives, right? The interregnum is over! A new age finally born! You could almost forget that even under a best-case scenario, with no mass outbreak of violence in America or far-right resurgence in Canada or Europe, but instead a concerted international engagement with climate change and inequality — even then, the odds remain heavily stacked against us.

Which is why we need hope more now than ever.

↞∿↠

Here's what didn't change on January 20: the Doomsday Clock, which remains set at one hundred seconds to midnight, closer to Armageddon than it — than we — have ever been. That's partly thanks to ongoing nuclear proliferation and geopolitical instability, but also because the Bulletin of the Atomic Scientists added climate change to its threat calculus in 2018, then COVID-19 in 2020. You have to wonder why they stopped there. Civilization has brought so many existential crises down upon itself — is there any need to list them? — that the prospect of nuclear annihilation has to compete for our attention.

It's only recently that any government on Earth has even acknowledged this general state of affairs, let alone tried to do something about it. This is as frustrating as it is understandable. In *Commanding Hope*, Thomas Homer-Dixon writes, "Anyone who grasps the severity of humanity's predicament . . . confronts an unforgiving conundrum, which I've come to call the *enough vs. feasible dilemma*. On one hand, changes that would be *enough* to make a real difference — that would genuinely reduce the danger humanity faces if they were implemented — don't appear to be *feasible*, in the sense that our societies aren't likely to implement them, because of existing political, economic, social, or technological roadblocks. On the other hand, changes that do currently appear feasible won't be enough by themselves." The battle over Canada's national carbon tax, vilified by Conservatives as an economic death blow and contested by four provinces all the way up to the Supreme Court, comes to mind.

Homer-Dixon describes himself as a complexity scientist. He's been studying the interplay between environmental collapse and society for decades, always with an eye for solutions that brings a rare note of optimism to this depressing subject. His previous book, *The Upside of Down*, argued that the decisive cause of the

Roman Empire's collapse was an energy shortage, with clear implications for the modern world; he closed it with a chapter on "catagenesis," his term for the way catastrophe gives birth to opportunity, which is where *Commanding Hope* picks up the thread. In Homer-Dixon's view, the institutional breakdown we're seeing all around us (exhibit A: U.S. democracy) is a counterintuitive source of hope. It's only in situations of radical breakdown, like a global pandemic or an armed insurrection, that radical policy changes become politically feasible. It's when things are at their most dangerous, Homer-Dixon argues, that hope makes the most sense.

Here's a story he tells about hope.

On September 12, 1961, a thirty-four-year-old activist from Connecticut named Stephanie May travelled to New York and began a hunger strike outside the Soviet Mission on East Sixty-Seventh Street. Two police officers threatened to arrest her for vagrancy the moment she arrived; only after she promised not to spend the night did they agree to let her stay. For the next six days, May occupied the sidewalk from eight a.m. to six p.m., living on water and broth and feeling, as she later wrote, "absolutely invisible, except to little children instructed not to look" at the resolute woman with a body-length sign strapped round her neck. It read *"Russia! Stop* Nuclear Testing!! Stop poisoning the air!"

The Soviet Union had announced it was abandoning the moratorium on atmospheric nuclear testing then in place between the world's two superpowers — a moratorium that Stephanie May played a prominent role in bringing about. History has largely overlooked her, but she was a key member of the Committee for a Sane Nuclear Policy. SANE, as it was known, was the most influential peace group of its time, with a membership that included Martin Luther King Jr. and Eleanor Roosevelt.

Over the previous decade, the great powers had measurably increased the entire atmosphere's radioactivity by detonating hundreds of nuclear warheads into sea and sky; cancer rates were spiking near test sites all over the world, especially in children. Now the moratorium May and so many others had campaigned tirelessly to achieve was about to be vaporized by a fresh barrage of tests. These were the days before Rachel Carson's *Silent Spring* and acid rain, long before Three Mile Island and Chernobyl. With Europe and North America entering the most prosperous period in human history, environmental contamination was a fringe concern and Americans' faith in their own government was matched only by their loathing for Russia's. A little nuclear fallout was a price the public seemed willing to pay to win the arms race.

Four days into May's protest, the press finally bit. "I'm not willing to crawl into a hole in the ground and accept nuclear destruction without a murmur," she told a reporter for the *New York Post*. "This is my murmur." Television crews from ABC, CBS, NBC, Universal Pictures, and others rushed to interview May the next day. Her poise and informed conviction inspired a wave of similar hunger strikes that spread across the country in the following weeks; May ended her own on its seventh day. Oblivious to the woman from Connecticut, the Soviets resumed their tests, and the United States followed suit.

In the following months, the Cuban Missile Crisis brought humanity to the brink of the nuclear abyss and triggered yet another round of demonstrative atmospheric tests. But one year after that, in October 1963, the Soviet Union, the United States, and the United Kingdom signed the Partial Nuclear Test Ban Treaty. They haven't detonated a nuclear bomb above ground since.

As Homer-Dixon argues, the reason Stephanie May's story matters isn't because she and her fellow activists won some total victory but precisely because they didn't. The threat of nuclear war has not abated — just the opposite — but our atmosphere's radioactivity *has* returned to normal over the past half century, as have cancer rates linked to fallout. For anyone paying attention in late 1961, such an achievement would have seemed somewhere between ludicrous and impossible. There was no rational cause to believe in it. You could only hope.

Sixty years later, as we face other lethal accumulations, it's good to be reminded that existential despair is neither new nor insurmountable. It's thanks to people like Stephanie May — and her daughter, Elizabeth, who served as leader of Canada's Green Party — that we have the luxury of contemplating why our odds keep getting worse. With that luxury comes the flip side of hope: the responsibility to act.

❮◇❯

"We inhabit, in ordinary daylight, a future that was unimaginably dark a few decades ago, when people found the end of the world easier to envision than the impending changes in everyday roles, thoughts, practices that not even the wildest science fiction anticipated."

So wrote Rebecca Solnit in *Hope in the Dark*, her response to the re-election of George W. Bush amid the ongoing horror of the Iraq War. Solnit was talking about the unlikely accomplishments of the anti-nuclear movement, which she, too, regarded as just one of the many reasons not to give in to despair at a political moment that was the opposite of the one we find ourselves in today. Back in 2004, instead of rejecting a Bush administration whose murderous incompetence was every bit as horrifying as Trump's, Americans resoundingly voted for four more years of mayhem.

Hope is a funny thing — it's when you need it most that it tends to be hardest to find. Maybe that's why we regard it with such suspicion. But for Solnit, as for Homer-Dixon, hope is a muscular emotion, inextricable from struggle. "Hope calls for action; action is impossible without hope," she reminds her readers. "Hope is an ax you break down doors with in an emergency." *Hope in the*

Dark serves as a literary tour of reasons to believe that a second Bush administration wasn't the only sign of the way the world was headed. There was the end of Apartheid and the collapse of the Soviet Union, the self-liberation of the entire South American continent, the steady march of women's rights and gay rights, and the notion, however poorly implemented, that nature has rights, too. These developments commanded hope.

"I say this to you not because I haven't noticed that the United States has strayed close to destroying itself and its purported values in pursuit of empire in the world and the eradication of democracy at home, that our civilization is close to destroying the very nature on which we depend — the oceans, the atmosphere, the uncounted species of plant and insect and bird," she writes. "I say it because I have noticed: wars will break out, the planet will heat up, species will die out, but how many, how hot, and what survives depends on whether we act. The future is dark, with a darkness as much of the womb as the grave."

The difficulty in talking about hope is that it's often used as a rhetorical sedative by people who say things like "never let a good crisis go to waste," or who offer some version of the false aphorism John F. Kennedy popularized about the Chinese symbol for "catastrophe" being the same as for "opportunity." That kind of hope is precisely what the NRA has in mind when it offers thoughts and prayers to the victims of gun violence. It's catnip for professional persuaders who would rather talk than change. And the thing is, it works, because who doesn't want their suffering to serve a higher purpose? When I listened to Justin Trudeau justify his prorogation of parliament in the summer of 2020 with the words, "As much as this pandemic has been an unexpected challenge, it is also an unprecedented opportunity," part of me wanted to forget everything but the possibility that the Liberals would finally do something radical.

Neither Solnit nor Homer-Dixon lets anyone off so easy — not me, not Trudeau, and least of all hope itself, which can only be trusted when it tells the whole truth. Their books, though different in tone (Homer-Dixon writes like a social scientist, Solnit like a prose poet), both read like a planetary version of the Stockdale Paradox, named after a man who survived seven years of torture as a Vietnamese prisoner of war. Talk about long interregnums. As James Stockdale put it, "You must never confuse faith that you will prevail in the end — which you can never afford to lose — with the discipline to confront the most brutal facts of your current reality."

↞↠

I was reading about hope when the Black Lives Matter protests broke out across the United States, and then the world, in the summer of 2020. The tragedies that sparked those protests were so heartbreaking and infuriating that they became a stand-in for the Trump administration itself. The breathtaking racism emanating from the White House, the impunity and the cruelty — these qualities weren't just reflecting the country's worst instincts, they were feeding and amplifying them. Definitely not a cause for hope.

But the sight of so many citizens marching, singing, filming, dancing, *rising up* together from all walks of life — now that was inspiring. It reminded me of an author I'd once heard on the CBC, a sociologist named Aladin El-Mafaalani, whose parents immigrated to Germany from Syria before he was born. El-Mafaalani is the author of *The Integration Paradox*, which he discussed in a speech at the Canadian embassy in Berlin.

"Compared to thirty or forty years ago," he said, "there is very clearly less discrimination in Germany, as well as in Canada, most

European countries, and the United States. But strangely enough, we discuss and fight over discrimination as if it had become worse." These fights, El-Mafaalani declared, were a counterintuitive sign of progress.

"When integration, inclusion, or equal opportunities are successfully implemented, they do not lead to a society which is more harmonious, or free from conflict. On the contrary. The central effect of successful integration is actually a higher potential for conflicts. In every case, more people will be sitting at the table and they all want a piece of the pie . . . How is this supposed to lead to fewer conflicts? This idea is either naive and romantic, in the sense of multicultural optimism. Or it is hegemonic, in the sense of expecting minorities to assimilate. Reality, however, looks different."

There's no question that four years of President Trump flushed a hideous amount of racism out of the shadows and fed it all the raw meat it could handle. Ordering a Muslim travel ban and calling Mexicans rapists and declaring Black Lives Matter protesters to be thugs and gangsters were not the kind of integration El-Mafaalani was talking about. But the backlash that Trump personified could also be understood as a perverse signal that things were heading in the right direction, for as El-Mafaalani noted, "Racism becomes especially extreme when it is directed against a group that is strong or becoming stronger, because that means some of the people sitting at the table are losing privileges."

In fact, El-Mafaalani regards conflict as an essential driver of human progress, so long as it is managed correctly. If you want to see what a society without conflict looks like, visit North Korea or Saudi Arabia. "Conflicts are energy," he said. "Energy for development, for improvement, and for progress. This energy can be something wonderful. However, energy can also take the form of a bomb. So conflicts are actually something which in and of themselves are neutral. It all depends on how one deals with them.

Conflicts which are managed in a constructive way can bring about wonderful things. Destructive ways of dealing with conflicts can lead to disaster. Wars are the result of conflict. But so are democracy, human rights, environmental protection, humanism, the social welfare state, the concept of an open society, and liberalism."

The conflicts arrayed before today's progressives are as harrowing as they've ever been. We are confronted by the most powerful forces in the world. It's not just fossil fuel companies we're up against but also the banks and shareholders who finance destructive industries, as well as the political power centres where the status quo resides. These are all democratic institutions. It was us who set them up. The closer you look, the more it appears that the thing we're really up against is ourselves.

To offer hope as a weapon in this fight runs the terrible risk of sounding earnest, which these days means insincere. Ours is the world of the corporate message and the public persona, a world of air-brushed influencers and pablum politicians whose sunny ways can win elections even as they make our eyes roll. The mirror image of all that PR is reflected in our private lives, where irony and cynicism have been gaining currency for as long as I can remember. When I'm among friends, without a camera, the savvy and sophisticated approach is to raise an eyebrow at any earnest promise.

We exalt irony over hope and praise the knowing chuckle because that feels like the smartest response. But lately I've been thinking, maybe it's just the easiest. The likeliest outcome for civilization over the next century may well be tragedy on a scale too large to contemplate. But that's not the only possibility.

In the final days of the 2020 U.S. election, I heard Alex O'Keefe, creative director for the Sunrise Movement, say, "Hope is a radical act of labour." The Sunrise Movement mobilized an army of youth activists to make the Green New Deal a household name and help elect some of the most progressive members in Congress. If they believe in hope, so can I.

Of course, it was also in reaction to organizations like Sunrise and movements like Black Lives Matter that many of those progressives in Congress came within a few inches of being murdered by Trump's berserker army. Today, as I write, there are 25,000 American troops assembled in Washington, D.C., while every state house in the country is under high alert against the possibility of a terror attack by members of the Republican base. The situation in Iraq that inspired Rebecca Solnit to write a whole book about hope has finally come home to roost. "Again and again, far stranger things happen than the end of the world," she observed, and lord have the past seventy-eight days ever proven her right.

Which isn't to say that worse, whatever that is, might not still lie ahead. It's just that right now, thanks to recent events, it's easy to grasp what all the wise people who write about struggle keep trying to remind us: Fundamental change may not always follow fundamental breakdown, but it never happens without it.

Wie man in den wald hineinruft, so schallt es heraus.
(As one calls into the forest, so it echoes back.)
— German folk saying

ONCE UPON A TIME IN DEUTSCHLAND

SOMETIMES THE PAST IS HARDER to imagine than the future. It doesn't even have to be that distant. How did we manage our social lives before cellphones? Did it really snow more when we were kids? Why did we let people smoke in bars? But if recent history can be difficult to plumb, the landscapes of bygone generations are murky as an underwater forest. You need the right apparatus to dive deep and get a sense of how it was before the dam burst.

Language itself is a kind of submarine. I'm thinking of a spectacular word: Vergangenheitsaufarbeitung. It translates roughly as "working off the past," and through its thick glass an entire black forest full of skeletons comes into view.

Vergangenheitsaufarbeitung takes me back to the first language I spoke, though I learned that particular word as an adult. We stopped speaking German at home soon after I started school, stranding me with a four-year-old's proficiency in my mother's tongue — decent accent, terrible vocabulary — and my only direct knowledge of the culture it expresses comes from my mother's character: unsentimentally blunt, yet besotted with the arts and addicted to dinner parties. Born in Germany in 1940, she started learning English when she moved to North America twenty-one years later. She was totally fluent by the time I arrived in 1976; I didn't even know she had an accent until I was ten, when my friend cracked a joke about it. Many things clicked into place in that moment. The punctuality, the point-blank style of asking questions, the permanent occupation of our stereo by Beethoven and Mozart and Bach — no Beatles or Bob Dylan in our house, nor anything else my friends' parents listened to.

This was the mid-1980s, when the two Germanys were at the height of atoning for the Holocaust. In 1985, West Germany's president, Richard von Weizsäcker, declared that May 8 would no longer be lamented as the day Germany lost the war, but from that year forward be officially celebrated as Liberation Day — commemorating the freedom not only of those in the camps but of all Germans. It was a major turning point, one of many on the endless road of working off the past. Forty years after Nazism was torn off their backs, Germans could stop thinking of themselves as the victims.

I was a child and oblivious to these distant developments, but I was already beginning to sense the way my mother's culture of origin came off as a caricature of humourless social rigidity. By high school, I knew very well that putting on a slapstick German accent was a good way to mock Nazis. *Ach, ze Jehmans ah coming! Und now ve dance!*

A funny thing, though, is how my mom becomes less stereotypically German when she's speaking German. In her native language, she is more nuanced and playful, more musical and spontaneous, than she is when speaking English. I see it in her face and glean it from the fragments I can understand of her quick and easy dialogue with the few German friends she has. I didn't realize this until long after I left home. Nor did I perceive the way her painfully direct speech in English could serve to conceal certain things, even — perhaps especially — from herself.

In any case, it's not that I felt guilty about my heritage or any more embarrassed by my parents than your average adolescent kid. But I wasn't about to sew a German flag onto my backpack, either.

↞∿↠

This is not a story about intergenerational trauma. I do, however, note that a gossamer cage I've been aware of for much of my privileged life is the defensive posture my mother always held on the subject of German culture. Mention Goebbels, she'll counter with Goethe; bring up Hitler, she'll parry with Merkel. This cage has not been opened by the celebration of diverse ancestries that has finally gone mainstream in North America. First-generation children of immigrants and Indigenous residential school survivors alike are shedding their childhood shame and venerating the cultures that produced them, publicly and proudly; rituals and cuisines and languages and stories that were previously cast into the melting pot are now enshrined as integral components of our cultural ecosystem. This wonderful turn of events was far too long in coming, and I offer sincere congratulations to everyone with the good fortune to not be descended from Nazis.

Not that I am — at least not directly — descended from Nazis. I just wish my forebears had tried to resist them. It hasn't helped that my mother could never explain how or why Germans fell under the spell of Hitler, an urgent question to my budding teenage mind and still a bit of a thing. There are, of course, whole libraries full of books that describe exactly how it happened, one brown shirt at a time. But some things you want to hear straight from the pferd's mouth, and it is unsatisfying to be told that it all happened before the pferd was born, and that there was no such thing as television or telephones in the village the pferd grew up in, let alone Google, and anyway the pferd and her family were too busy surviving the hard postwar years to interrogate the past.

The Third Reich was an abomination, no argument from her there. But in my mother's eyes it was also an aberration, one her country had atoned for, and that was that. Now I was coming of age in a country where dawning awareness of the Indian Act's myriad atrocities were frequently met with shrugs of *It wasn't me who did that*; on top of that, evidence was mounting that our daily habits were actively destroying the planet's biosphere, without anyone seeming much bothered. Among the first news events I became acutely aware of was the extinction of Atlantic cod; we just . . . ate them all. I was developing strong, if inarticulate, feelings about the relationship between indifference and complicity. I felt implicated in everything, powerless to prevent anything. Most days, I still do.

The most my mom could tell me about the era predating her birth was that her parents did, yes, vote for Hitler, but only because he promised a monthly government cheque for each of their four children. His was the National Socialist Party, after all. By the time my mother could speak, her family had developed an impenetrable silence on all questions of the past. School was no better. My mom entered the work force after grade eight, but even if her parents could have afforded a proper education she wouldn't have learned anything about Nazism, because Germany's history classes had a 1933–1945 hole in their curriculum back then. She recalls a vague awareness that certain houses where Jewish people once lived were no longer lived in by Jewish people. My grandfather, a gentle soul, surprised her once with a stream of harsh words for a neighbour who'd taken part in something called Kristallnacht. And there was the day when my mother answered a knock on the door and met a woman who introduced herself as Hannah and asked if her mother was home. My grandmother emerged from the kitchen, and when she saw Hannah she rushed forward and locked her in a tight embrace and cried and cried, and cried. A childhood friend, Hannah had escaped to Britain. She lived there still. After that and for the rest of her life, Oma visited London periodically to spend time with the friend she thought she'd lost.

So, there were hints. There were inklings. But these were never unpacked, as we would say today. It wasn't until my mother moved to North America in 1961 that she learned about the Holocaust.

Now I'm the parent. Soon it will be my turn to answer questions about things that could have been prevented. May I suggest, dear daughter, that this is the wrong line of inquiry. I know you'll probably disagree, as I did, but hear me out. The thing we should be asking isn't how an entire culture slips beneath the waves but how it returns to the surface.

How do you work off the past?

"Vergangenheitsaufarbeitung was one of the first words I added to my vocabulary," writes Susan Neiman in her 2019 book, *Learning from the Germans*. An American Jewish philosopher who grew up in the deep South, Neiman moved to Berlin as an adult in 2005; she's been there ever since, directing the Einstein Forum and interviewing some of the most interesting people in the country. In *Learning from the Germans*, she charts Germany's postwar journey of coming to terms with its sins; the point, as the title implies, is to see what other nations might learn about dealing with their own buried crimes.

"Working off Germany's criminal past was not an academic exercise," she writes, "it was too intimate for that. It meant confronting parents and teachers and calling their authority rotten."

If that's something every generation does in every country, in Germany the repudiation rose in proportion to the rot. (In West Germany, that is — East Germany defined itself from the beginning as an antifascist state governed by leaders of the former resistance; they remembered that the first people sent to concentration camps weren't Jews but communists.)

For the first fifteen years after the war, Neiman writes, West Germans "confined their memories to the end of the war: the bombings, the losses, the hunger. The idea that any of this might be considered deserved punishment for starting and supporting the deadliest war in human history crossed hardly a mind. The nation stewed in its own pain, and devoted itself to cleaning up the cracked brick and broken concrete strewn through its cities. It took decades before there was much interest in tackling the moral ruins."

By the time that second task began, my mother had departed the continent. For better or worse, she never looked back. So when Neiman writes, "Because their parents could not mourn, acknowledge responsibility, or even speak about the war, the next generation was damned to express it," I can't help feeling that sentiment got bogged down in the transatlantic voyage and took a generation to catch up. Time and distance have watered down the potency of my forebears' unfathomable sins, but when I look at the storm clouds gathered on our century's horizon, learning from the Germans feels more urgent than ever.

↞∿↠

When I was three, my mom took my brother and me to Darfeld, the Westphalian village where she was born and raised, to meet her parents. All I remember is falling into a patch of nettles. My grandparents died before I could see them again. They left behind some black-and-white photographs of a time before people automatically smiled for the camera, and my mother's stories. Opa was a happy-go-lucky hustler who never let the family go hungry, but never got them much above the poverty line, either. Well, that was as good as it got for most of the first half of the twentieth century in Germany, aside from a few good years in Berlin. Oma was a long-suffering soul who endured the usual outrages inflicted on women everywhere back then, brutally amplified by the war and its aftermath of penury. But she wasn't widowed by the war (my grandfather had a shoulder injury that saved him from the draft), nor did she lose any children (her only son was born in 1942), and that was more than most German women of her generation could say.

My mother has two memories of the war. One is of the whole family racing out of the house during an air raid and rushing toward an underground shelter that they couldn't reach in time; instead they huddled under a pine tree while planes roared low overhead. The other is of peering over the windowsill at Allied bombers rumbling over the countryside, too far away to trigger the town's alarm but close enough to hear. I sometimes think of those things when my own daughter — now the age my mom was then — shouts and points at an overhead jet.

But overall my mom remembers her childhood in surprisingly happy terms. Darfeld had many things going for it then that most of Germany did not: as a rural backwater, it was spared the full fury of the Allied assault that reduced cities to rubble and killed a million civilians. The farmland surrounding Darfeld also kept at bay the famine that gripped the industrial heartland right after the war. And in a final saving grace, Darfeld had the excellent fortune of being occupied by the bonny English rather than the savage Russians, tales of whose rape and pillage make my mother shudder to this day (of course, no mention of Russian barbarity is complete without acknowledging that which preceded their own: some twenty-six million Russians were killed during the war, most of them by Germans). The occupying Tommies stationed in Darfeld were kind. They handed out candies to kids. My mom remembers running toward them whenever a patrol cruised by.

There was a small medieval castle on the edge of Darfeld. For a few years following the war, it became an impromptu refugee camp for Czech and Polish volksdeutsche expelled from their countries in the wake of Germany's surrender. The refugees spoke a thickly accented German peppered with strange words. Nobody in town (except the castle lord, forced to convert his stables into human lodgings) gave the new arrivals much thought at the time; on the scale of crazy things happening throughout Europe in the 1940s, a few nomads settling down between the hay bales didn't score all that high. They had never rated a mention when I queried my mom about her childhood. Not until I learned for myself about Germany's postwar refugee crisis and asked her if Darfeld had been affected did she remember her volksdeutsche neighbours. Even then she had to scratch her head and think, giving me the impression that she was only just now piecing together what had brought them to Darfeld.

Which is funny, but also understandable. Funny, because those ragged volksdeutsche represented the biggest single act of ethnic cleansing in history: the purge of fourteen million ethnic Germans from Eastern Europe in the wake of Nazi occupation, something we would notice if it happened today. Understandable, because of the trauma that the people of Darfeld themselves were coping with. Before they began working off the past, Germans did just the opposite: They ignored the past and didn't get too curious about unusual things in the present.

In any case, I only mention that historic movement of the masses because one of its offspring is me.

<↔>

About three million of the volksdeutsche who wound up in Germany came from Czechoslovakia's Sudetenland. Among them were several Kopeckys, a hybrid family like so many of the volksdeutsche, with a Slavic surname but a strong sense of German identity. My dad was born in the Sudetenland in 1932. He remembers his parents switching from German to Czech when they didn't want the kids to understand what they were saying.

He would have learned Czech eventually, but his father moved the family to the United States in 1934. He'd bought some farmland in Iowa during a previous journey, and with the Great Depression threatening his investment, he decided to go back and take care of things himself. Opa Iowa, as we still call the man I never met, brought his young family with him, intending it to be a brief sojourn. But the Third Reich and its aftermath put an end to all dreams of return to the Old World, turning my father and his siblings into permanent Americans.

My paternal Opa and Oma and their kids absorbed American propaganda, not the Third Reich's. My father came into adulthood understanding America to be the global good guy, the country that defeated Hitler and was now going to stand up to Stalin. This stance was slightly complicated by the fact that several of his uncles who'd stayed in Czechoslovakia had eagerly joined the German army — they were Nazis — but, by 1946, they were either dead or broken. With the war over, the surviving Kopeckys in Czechoslovakia got the same knock on their door as three million of their Sudeten volksdeutsche neighbours. A train was leaving for Germany in four hours. They could board it or die. They took the train. A few wound up in a Westphalian village called Darfeld, where they moved into the stables behind a castle on the edge of town.

My father paid them a visit in early 1961, and that's when Karl met Rita. Three years later, my parents were married and living in

Edmonton, where my dad was teaching chemistry at the University of Alberta.

The geopolitical tempests that brought my parents together had finally subsided. My dad worked while my mom studied and graduated high school, then university, and then she became a teacher in the Catholic school system. They had two children. Then they divorced. My dad moved five blocks away, and I grew up skipping between two houses, as oblivious to the emotional gulf that lay between them as I was to the tens of millions of violent deaths that cleared the way for my placid middle-class upbringing.

~◇~

My parents were doubly fortunate. First they emerged unscathed from a perilous childhood. Then they entered adulthood at the start of the most prosperous period of economic growth in human history — a one-time explosion of distributed wealth where rising-tide economics converged with trickle-down theory to soak three-quarters of the population in well-being.

The opportunity landscape of postwar Europe and North America (for white people) resembled a forest that's just been devastated by fire: Everything blackened and burned, but underneath the soil the roots of society remained intact. New seedlings no longer had to compete for sunlight; the canopy was gone. American money — the Marshall Plan, the World Bank, the International Monetary Fund — fertilized the soil on both continents. But North America, with its unparalleled industrial infrastructure completely intact, had the clear advantage. An older, deadlier fire had already swept through and cleared over 90 percent of the original inhabitants before burning itself out. The victims of that genocide left a different kind of treasure behind than that of Europe's Jews, and it was just as eagerly plundered. Uncut woods and undammed rivers, mountains full of coal and minerals, great lakes of oil, buffalo pastures cleared of game and ready for the plough, everywhere space, space, space. In 1950, the year my father turned eighteen, North America was absolutely loaded, even if he wasn't.

By the 1970s, the physical landscape of this continent was badly denuded, but the forest of society had regrown. No sign of the great fire anymore, everything green and growing. The middle class contained 61 percent of Americans in 1971, with another 14 percent in the upper class (statistics that mask the breathtaking exclusion of Black and Indigenous citizens, among others). My mom, born in a war zone and growing up penniless and uneducated, taught elementary school for thirty years and then retired at sixty with a comfortable pension and a house she owned outright in a fine neighbourhood (where a literal forest had stood one hundred years before). Imagine. By the 1990s, while I wandered through high school and into university, things like publicly funded K-12 education and health care, three healthy meals a day, minimum wages, and at least one car per family seemed as ordinary to me as leaves on a tree. It was the natural order of things that every passing year would bring an increase in abundance: more income, better technology, greater varieties of food and music and wine and travel options. We were rich. We wanted to be richer.

While everyone was looking up, the carpet began slipping out from under us. Some of this was self-inflicted, especially in the United States. No single statistic can capture the full measure of the way America shredded its social safety net. But one that comes close is the top marginal tax rate, which rose above 80 percent in the United States in 1940, and kept rising to reach 92 percent by 1952. It stayed above 90 percent for the next decade, then began its libertarian descent; by 1990, it had dropped to 28 percent. The Cold War was over and capitalism no longer had to prove how much better it was at providing for the masses. The 1990s became a decade-long victory lap for America and the "Western" world, a sprint to riches that erased any remaining distinction between freedom and greed. The stock market became a rocket that escaped the gravity of the social contract. The natural world collapsed far beneath its orbit, unnoticed. Climate change, the subject of mounting bipartisan American concern throughout the 1980s, ceased to matter. The internet was born. Such was the giddiness of the moment that a brilliant journalist named Francis Fukuyama could, in 1992, write a book arguing that the end of history had arrived, and instead of being laughed out of the room, he was celebrated as a prophet.

Taxes can be raised again, safety nets rewoven, stock markets regulated, social contracts renegotiated. Ecosystems, however, do have points of no return: Ancient forests and freshwater, minerals and fish, top soil and atmospheric carbon concentrations are all approaching thresholds that should scare us just as much as Nazis. Cod: gone. Salmon: going. Forty percent of the ice in Glacier National Park: gone. One thing there's a surprising amount left of: oil.

By 2020, our trajectory has become the opposite of what our parents lived. They would never have less than what they were born with; we will never have more. Own your own house in the heart of a city? Retire at sixty? Live off your savings? Go on. Our roots have drained the aquifer; the forest is drying out once more; there have been a few economic tempests in the course of our lives, but no all-consuming cataclysm for eighty years.

My father is approaching ninety, my mother now past eighty. The living memory of a world war — of the essential fragility of any good life — will soon be gone from this earth. With it, I sometimes fear, whatever vestiges of restraint that have held us back from doing it all over again.

But maybe war's not the thing we should be worried about. Eighty years ago, the fight was against oppression. Today the threat is freedom. The spectre of violent suppression has been replaced by the consequence of excess liberty — freedom to eat anything, go anywhere, hoard enough money to feed a billion people. You can put whatever you want in the air, take whatever you want from the ground, surround the Galapagos with fishing boats, dam every last river in the world — two-thirds of them so far, and counting.

Who would have thought, in 1940, that freedom would one day become a problem?

After my dad retired from a lifetime of workaholism, he began reading the news more carefully and became aghast at what the United States had turned into while he wasn't looking. This was around the time Bush launched the War on Terror and terrorized Iraq. Karl Kopecky took it as a personal betrayal. The America he grew up in was a planetary good guy, or so he'd been raised to believe. But the more he read, the more he decided that none of this started on 9/11, and that the official version of U.S. history was every bit as fictional as the Bush administration's rationale for invading Iraq. Defeating the Nazis had cloaked his adopted country in a heroic narrative that buried all mention of its own original sins, equal in magnitude to the Holocaust: the genocide of Indigenous Peoples and the enslavement of millions of Black people. Today if you ask him — actually you don't need to ask him — when it all began to go wrong, he won't point to Bush or Reagan or the Vietnam War or the Old South or the Wild West, but all the way to Plymouth Rock. He's been working off the past ever since.

The discovery of all those skeletons in America's closet was the closest thing to a religious awakening an atheist like my father could experience. At the age of seventy, he became a born-again radical: a left-wing vegan anti-war climate change activist who ranted on behalf of every underdog and minority group on the planet. He started wearing a black armband every time he left the house to represent the dead in Iraq. He became the guy you didn't want to get cornered by at a dinner party. He ordered suitcases full of "Bush Bucks," fake American dollars that pictured Dubya dripping in blood and oil against a backdrop of planes flying into the Twin Towers, and handed them out gleefully to strangers and friends alike.

This was awkward enough in Edmonton, where most of his friends shared a less feverish version of his politics. But it became positively fürchterlich on his annual visits to the family farm in Iowa each fall.

The trouble always started at the border crossing. In 2005, when a customs agent asked him about the purpose of his visit, my dad replied that he was looking for weapons of mass destruction, because Bush had promised to find them in Iraq but still hadn't found them; perhaps they were in America? Somehow, after several hours of questioning, he was allowed through, and the crusade pressed on.

My uncle and aunt and several extended relatives, being Iowan farmers, hewed to a different philosophy. They were devout Christians who worked hard, loved their neighbours, didn't touch alcohol, and watched Fox News seven days a week. We used to visit them every summer when I was a kid, in the days before Fox News. I loved those trips to the farm. Uncle Rudy was a delightful prankster with a slow unflappable drawl, patient as the weather; Aunt Jeanine was his eternally exasperated wife who took care of everyone and everything that wasn't a piece of farm equipment; their children, my cousins, were a few years older than me and had a rare combination of playfulness and competence — clowns one minute, they'd fix a broken tractor the next, then get back to tickling me. I adored them all.

Those summer trips tapered off by the time I reached high school, but my father kept up his annual pilgrimage. I heard through him about their slow poisoning by American Conservative media, in particular the daily drip of Fox News. Thanks to that toxic flow, many — not all — of my Iowan brethren had come to regard American cities as dangerous, immoral places, full of welfare queens and carjacking bandits and socialist whiners who didn't believe in God but liked to lecture redneck farmers like my cousins — an impression my atheist, city-slicker father did little to dispel.

For years, my aunt and uncle put up with his proselytizing. Then one Sunday in September, while they were all in church, my dad went through the church parking lot and slipped a pamphlet under every car's windshield wiper. The pamphlet explained that the Republican Party was a terrorist organization driving the world straight to hell.

It didn't take a genius to figure out who the culprit was. When the family confronted my father that evening at supper, he acknowledged the crime and doubled down: He rose to his feet and declared they'd all turned into Nazis, embellishing the statement with a stiff-armed Nazi salute and a heil Bush!

That one got him excommunicated. His banishment was no doubt a relief for Iowa, but it was tragic for my dad. Iowa was the closest thing he had to a spiritual anchor. Despite his urban life and recent political awakening, he remained a farmer at heart. Socially conservative (he made it through the 1960s without smoking a joint), married to his job (a prime factor in the breakdown of his other marriage), my old man always fit in better at an Iowa county fair than the university's faculty club, and to this day he opens our every conversation with a detailed dissection of the weather and its likely impact on farmers. He suffered through five years of exile from Iowa; finally my brother and I, who kept in loose touch with our cousins, negotiated a rapprochement. He would be allowed to return on the condition that politics never came up again. He stuck to the terms of the peace treaty, even through the Trump years.

I stopped by the farm on a trip through the U.S. in December 2016, during the interregnum between Trump's election and inauguration. My aunt and uncle received me as warmly as ever. They had a clock in the kitchen with Obama's face on it, counting down the remaining days and hours of his presidency. Conscious of the rules, I didn't bring up politics, but a friend who joined us for breakfast one morning did.

"That man is the anti-Christ," he said, pointing cheerfully at the Obama countdown clock.

I asked, genuinely curious, why he thought so. Obama, he replied, was on the record stating that the founding fathers were Muslim and that America is not a Christian nation. The president had spent $85 million of taxpayer money on vacations so far, and that wasn't counting this coming Christmas. Nobody really knew for sure where he was born, and did I know that Obama refused to release his college records? There was no proof he'd actually gone to Yale. At the very least, it was clear he had something to hide.

Besides all of which, he'd let thousands of Syrians into America who claimed to be refugees. "When you think of a refugee, what do you picture? Desperate, hungry people, mostly women and children, right? But you look who we're letting in and it's strong, angry young men in their twenties and thirties!" I said my understanding was the United States did background checks on everyone seeking asylum and that there hadn't been a single terrorist attack caused by a Syrian in America. "Not yet," came the reply, "but look at what happened in Germany." By coincidence, we were speaking one day after a man from Tunisia drove a truck into a Christmas market in Berlin, killing twelve people and injuring another fifty-six; the attack was all over the news just then. "Now, Donald Trump might not have been anyone's first choice, but he knows how to run a business, and a country has to be run like a business. Liberals keep saying he's racist, but anyone could see the man was happy to work with anyone, Latinos, Blacks, you name it."

"Besides," said my aunt, "he'll be surrounded by good people. They'll make sure of that. What I don't like is the way people are protesting now that he's won, breaking windows and being belligerent. We didn't do that when we lost. You got to accept the result."

Sometime around this point, my uncle Rudy, who hates politics and was clearly embarrassed by the conversation, interjected. "Hey, let's talk about the weather! Isn't it supposed to rain tomorrow?" I said sure and invited them to my wedding. My partner and I were getting married the next summer, and I was hoping they would join us.

My aunt and uncle looked at each other, touched but a bit uneasy. "My knee doesn't let me get around too well," Rudy said. Jeanine brought up a news story she'd seen the other day, about a grandmother who visited her children in some city or other. "She was driving with her grandson in the back seat, but she got lost, and when she reached a stop sign, she looked at a map to figure out where she was going. Well, the man in the car behind her got mad cuz she'd stopped so long. He got out of his car and went up and shot her grandbaby! Just cuz she stopped so long! That's the kind of thing going on out there."

I promised she wouldn't have to worry about that in Canada. The wedding would be on a small island, far from any cities. Eight months later, my aunt and uncle made the trip and danced beside me on my wedding night.

As for the friend who'd so revelled in the sins of Obama, he travels with me now in spirit. I think often of the sincerity of his convictions. Before we'd parted ways that day, he told me he used to hate the Electoral College, but then he "did a little research" that taught him how essential it was to American democracy. If it wasn't for the college, then Chicago and the coasts would elect every president. "You gotta know your history," he said, "otherwise you're bound to repeat it."

My mother didn't want to talk about her history; my dad wouldn't shut up about his. I doubt this came up in their marriage or subsequent divorce, but today it strikes me as a powerful metaphor for their relationship: an oppositeness that brought them together before it drove them apart.

I sometimes see my own life as a search for the middle ground they must once have occupied together. I became a writer because I loved stories, and perhaps the stories I naturally gravitated to — stories about social justice and ecological destruction — were influenced by an unconscious sense of debt. The need to atone for my ancestry. Of course, that's a nice way to think about myself. If I sharpen the lens further inward, more unpleasant things come into view. A lifelong fear of conflict, giving rise to a trait Nietzsche might have derided as a will to please, takes shape. I see that another allure of writing has always been the safety of the solitary room. It's easy to hide behind the written word; writing saves me from having to challenge people face to face, or think on my feet, or manage my emotions before an audience.

On good days, my aversion to conflict is entirely Canadian, that polite and passive demeanour, the clichéd apologetic reflex. On good days, I'm repudiating the extremes to which my ancestors took conflict. But on bad days, the polite demeanour is just my way of stifling a more dangerous trait, one both my parents expressed in their opposite ways, which is an all too German sense of superiority.

Germany's culture of excellence in everything from arts to autos, when untethered from love and humility, can explode into a culture of domination. That's the will to power Nietzsche romanticized, the one Hitler got the whole country drunk on. Germans don't have a monopoly on this — in English, it's known as white supremacy — but we took it further in more recent memory than any other culture. This is what Thomas Mann was thinking about when he said, rather understatedly, "It would be hard to find a politician, a statesman, who accomplished great things without having to ask himself afterwards whether he could still regard himself as a decent individual." A line from Goethe's "Erl King" that my mother likes to quote puts it more succinctly: "If you're not willing, I'll use force."

This is the fever that lures certain environmentalists into the woods of eco-fascism (the Nazis were great lovers of nature and would have happily exterminated whole nations for the sake of preserving wilderness). I feel it, too, somewhere in the depths, and fear it. It's the one thing my people, so famous for loving control, can't control. That's why, for two generations after the war, the Germans who stayed in Germany kept a tight lid on their emotions, and above all their pride. Don't let your passion run away with you; do *not* follow the courage of your convictions — remember where that got us. It's why they didn't dare to wave the flag at soccer matches until the twentieth century was almost done. And it's at least part of the reason why a German kid who grew up seven thousand kilometres, two generations, and one language away from the Third Reich spends a lot of time thinking about the tensions between pride, humility, and confidence.

↤ᨆ→

"Literature has always been sure of one ally: the future," wrote Günter Grass in "The Destruction of Mankind Has Begun." "It has outlived absolute rulers, theological and ideological dogmas, dictatorship after dictatorship." A novelist, essayist, political activist, and Nobel laureate, Günter Grass did more than almost any other German to help his country work off the past. He shot to fame in 1959 with his first novel, *The Tin Drum*, which depicts the rise of Nazism through the eyes of a child who decides to stop growing at the age of three and develops the magical ability to cut glass with his voice. That book's outlandish fabulism exposed the fundamental absurdity of Germany's dictatorship, and the irony that its authority depended on being taken seriously. The satirical force of Grass's magical realism stretched well beyond Germany and soon inspired a young novelist named Salman Rushdie to try something similar. But outside of Grass's fiction there was nothing magical about his realism.

"Born in 1927," he wrote, "I belong to a generation that although it may not have directly participated in the German crime — the genocide of six million Jews — bears to this day the responsibility for it and is neither able nor willing to forget it." Like most German boys his age, Grass was an enthusiastic member of the Hitler Youth. At seventeen, in the war's final desperate weeks, he was called up to fight against the Russians ploughing straight to Berlin; almost all the men in his division died, and Grass was extraordinarily lucky to have survived. He never stopped wondering what monstrosities he might have committed had the war lasted a little longer, or if he'd been old enough to join it sooner.

But it didn't, and he wasn't, and he became a writer instead of a murderer. While his countrymen reconstructed Germany, Grass reconstructed German. I don't want to say he cleansed it, because that's the kind of word Nazis might have used, but certainly Nazism corrupted language, and through language, thought. Grass's writing became a kind of literary anti-corruption campaign, a ceaseless quest to "take the goose step out of German," as he put it. In his hands, the stern vocabulary of violent conquest and glory gave way to a lexicon of empathy, humour, and wonder. He was hardly alone in this — hero worship is another Nazi habit that I must pull myself back from. It's also crucial to acknowledge that Grass, the great scourge of German silence, had secrets of his own: Not until 2006, when he published his memoir *Peeling the Onion*, did he admit that he'd been called to active duty in the final days of the war (until then, he'd let on that the furthest his Hitler Youth membership had taken him was manning an anti-aircraft cannon in Danzig). But few would deny that Grass was a leading member of the generation of German writers who "tore their language down and rebuilt it anew," as Salman Rushdie observed.

By the 1970s, time and novels like *The Tin Drum* had begun to put the rise of Nazism into the perspective we see it through today. This was the pivotal decade when Germans stopped hoping it could all remain unspoken. Grass became a speechwriter for Chancellor Willy Brandt, a former leader of the resistance who spent the war in exile. It was Brandt who paid Germany's first post-war state visit to Poland, in 1970. In a gesture that instantly became famous, he fell on his knees before a monument to the Warsaw Ghetto Uprising, keeping silent for thirty seconds. "A speechless event that left nothing unsaid," wrote Grass, who was there. Many older Germans denounced Brandt as a traitor at the time. But that image of a German chancellor kneeling, head bowed, at the feet of his nation's shame, moved millions and marked a turning point. Vergangenheitsaufarbeitung had its first West German symbol. The era of atonement had begun.

By 1970, Grass could add the present and future tense to his prose. "Writers have recently begun to take an interest in conflicts that are no longer predominantly products of the war and of the postwar period," he said that year, at Germany's inaugural Writers Conference in Stuttgart. "Peace, that still-unexplored territory, confronts us with unaccustomed tasks. No more decisive battles fraught with history, no more Gotterdammerung or invocations of ultimate goals, no Weltgeist on horseback. Instead, such problems as . . . the new irrationality of technological mysticism, aggressions spawned by the mass media, terrorism, productivity, the parallel increase of pollution and prosperity are beginning to find their authors.

"Who will write about the slow death of Lake Constance? About the degradation and defence of the environment, the crisis in the educational system of a society dedicated to frenetic achievement, about the surfeit that comes of glut? What writers will give these issues form and content. . . . what new readers will they attract?"

Grass was well aware that these new stories were no more guaranteed a happy ending than the ones that came before. When he spoke about the future having always been the writer's ally in "The Destruction of Mankind Has Begun," he was speaking of the past; his point was that writers could no longer take posterity for granted, thanks to a threat even greater than the Nazis. "Our present makes the future questionable and in many ways unthinkable, for our present produces . . . poverty, hunger, polluted air, polluted bodies of water, forests destroyed by acid rain or deforestation, arsenals that seem to pile up of their own accord and are capable of destroying mankind many times over."

It would be an exaggeration to call Günter Grass an environmentalist. But I don't think it's a coincidence that, in the process of renovating his own language, Grass unwittingly contributed to the new language of environmentalism. Nor should anyone be surprised that one of the greatest postwar contributions to humanity of his guilt-ridden nation would be the development of renewable energy technology, in particular solar power, at the behest of a Green Party that has more influence on German politics than those in any other nation.

↔

When I was a teenager, I often heard adults say some version of *We're sorry you're inheriting this fucked-up world, but we believe in your capacity to fix it.* Now I'm a grown-up, and I hear people like me say the same thing to Greta Thunberg's generation, and they look just as consoled by it as we did (though they're much better at organizing marches). I admire Thunberg more than words can say, but one day she'll turn forty, and there will still be problems.

When I think about that cycle of inheritance and transmission, I think about the questions my daughter will, I hope and fear, eventually ask me. Perhaps she'll ask about the rise of Donald Trump. Perhaps she'll want to know how we could still fly planes around the world for fun, or eat beef and wild salmon, or go through so much plastic, in spite of all we knew.

Such thoughts of future reckoning summon Günter Grass's essay, "A Father's Difficulties in Explaining Auschwitz to His Children":

> When I try to explain Auschwitz to my children today — and they do demand an explanation, coolly suspecting something and openly curious: "So what was that about?" — my explanations soon become long-winded and circuitous and complicated, and I stop making sense altogether. It's as if with each step I take toward Auschwitz I have to take two steps back. Again and again in the middle of a halfway adequate explanation I see other reasons that have to be mentioned. And before I even get to those reasons, yet more reasons turn up: It was we, too. It wasn't something we wanted. But whatever we did and said and wrote led indirectly to that place called Auschwitz.

These words resonate with me today, but I know I must be cautious. You can't compare environmental collapse to the Holocaust, because you can't compare anything to the Holocaust, and for good reason. No amount of non-human death is morally equivalent to the extermination of six million people, including children. People say, *Well, climate change will soon be killing more humans than*

Hitler ever did. That may become true, but it isn't yet. Even if it does come to pass, death by unintended consequence isn't morally comparable to homicide; in legal terms, this is the difference between criminal negligence and first-degree murder. It's true that Exxon knew, as the hashtag goes, and so did countless others; true that some people have cynically (that is, profitably) dedicated their lives to delaying climate action. The Rex Tillersons of the world, and the institutions they've led, should be held accountable for the public impact of their misinformation campaigns in the same way Big Tobacco and the opioid makers have been called to account for theirs. But none of them compares to Hitler and his architects.

Climate change deniers don't put children into ovens. What's more, coal and oil and natural gas have performed a service for humanity that Nazis could never lay claim to. Fossil fuels may be ravaging the planet, but they've also delivered a level of prosperity that our species has never before known. By contrast, not a single good thing came of Nazism. Even those "Aryans" who supposedly prospered under its regime for thirteen years were themselves poisoned and debased; they might have led materially wealthy lives, but there was nothing to envy about the psychological price they paid for dominance. "It was a militant slave mentality," wrote Thomas Mann, who described National Socialism's paradoxical pursuit of "world enslavement by a people themselves enslaved at home."

Now that we know the damage fossil fuels are causing, the crime of continuing to use them is more severe. The stakes are so high, the scale of change needed so immense, that many activists make a direct comparison to World War II: We must wage war on climate change, because the threat it poses is every bit as drastic as the threat once posed by Hitler. Just as we retooled our entire economies in 1940, we must retool them today. Nationalize whole industries to produce the necessary hardware (instead of fighter planes and battle ships, wind turbines and photovoltaic cells), ration gasoline, overhaul the tax code the way Canada did in 1940, so that 100 percent of a corporation's "excess wealth" goes into public coffers to fund the war effort.

Anyone who grasps how hard it will be to keep the climate from warming more than two degrees has to sympathize with that rhetoric. Humans respond to war. We know how to mobilize for war. Thousands of years of war have steeped our languages in a vocabulary of violence, so ingrained it takes a conscious effort to avoid. We declare war on poverty, drugs, terror; we battle addiction; we beat, or are beaten by, cancer; we crush, we slay, we kill. We fight for what we believe in.

There are times when you do have to fight. Life *is* bloody, and cruel, and violent. But it isn't only that. We don't have to turn all challenges into enemies, every conflict into war. That's often a choice. There are usually others. Military metaphors have a way of shrouding those alternatives.

When I think of how Germany worked off its past, it strikes me that a big part of the process was to pacify the language. The pursuit of "peace, that still-unexplored territory," and its unaccustomed tasks, was largely a project of disarming the way people spoke, and wrote, and ultimately thought.

In April 2020, when Donald Trump was declaring war on COVID-19, Germany's president, Frank-Walter Steinmeier, objected to that framing. "This pandemic is not a war," Steinmeier said. "Nations are not opposed against other nations, soldiers against other soldiers. It is a test of our humanity."

The same is surely true of our quest to reconcile the way we live with the ecological requirements of the planet. It's hard to imagine that future peace, let alone how we'll overcome our differences to achieve it. But there are hints and precedents in all our lineages. By now, the relationship between truth and reconciliation is very well established. We know which comes first. Germany's example is unique, but if it strikes you as exceptional, try looking harder at your own.

What is any ocean but a multitude of drops?
— David Mitchell, *Cloud Atlas*

EVERY LITTLE THING

SET AGAINST THE MAGNITUDE OF this crisis, our inescapable smallness.

There's just no way around it: I can swear off the internal combustion engine, renounce meat, cover my roof in solar panels, touch no plastic, and plant a hundred trees a day for the rest of my life, and none of it will bring the atmospheric concentration of carbon dioxide any closer to 350 parts per million, nor ease the world's biodiversity crisis. Those things will only happen when hundreds of millions of people change their consumption habits. And *that* will only happen when governments pass the right laws.

That's why focusing on government and industry, as opposed to individual behaviour, has become the default position of the environmental movement. For too many years, governments and corporations alike urged us puny citizens to do our part without bothering to do theirs. Reduce, reuse, recycle, and shop green while you're at it; refrains like these have allowed the big players of industrial society to shift their ecological responsibility onto our frail shoulders.

They're still doing it. In fall 2020, Shell Canada launched its "Drive Carbon Neutral" program, kindly giving drivers the option of paying two extra cents per litre of gasoline on carbon offsets, in the form of a forest conservation project. This means, if Shell's done the math right (a palatial *if*, but let's grant it for the sake of argument), that for every molecule of carbon dioxide released through your exhaust, an equal number of molecules will be inhaled by a tree whose life was saved by your two cents per litre. Never mind that there's no contingency plan should the forests Shell has chosen to sequester your carbon burn down because of climate change. Never mind that offsets like these encourage people to use gasoline and thus slow the transition to electric vehicles and public transportation. And never mind that Shell spent $49 million in 2019 lobbying governments not to enact climate legislation, second only to BP among the world's oil and gas companies — BP being the company that introduced the very notion of a carbon footprint into our lexicon through a brilliantly devious marketing campaign in 2000, designed to make the public think of carbon emissions as a personal responsibility. Despite these objections, one might still find it in one's credulous heart to suppose Shell was trying to do the right thing. Where the line got irreparably crossed, however, was when Canada's environment minister, Jonathan Wilkinson, used his office to advertise the program (that is, advertise for Shell) and urge Canadians to use it. "It's forward thinking initiatives like this that will help Canada reach its goal of net-zero emissions by 2050," Wilkinson tweeted out to his followers, provoking the only good thing to come out of this embarrassing affair, which was a flurry of entertaining replies.

To be clear: It is the responsibility of high-office holders to put policies in place that will facilitate positive systemic change; it's also their responsibility to hold large polluting companies to account. If Canada were anywhere close to meeting its own climate targets, Wilkinson's support of Shell's advertising ploy might have been forgivable. But we're not. And until we are, the only message we should be hearing from politicians like Wilkinson or companies like Shell is what *they* are doing to decarbonize.

That said, let's not forget who's been choosing our politicians while the world burns down around us.

Nothing illuminates the vital importance of individual behaviour like an election. We know, objectively, that no single ballot is remotely likely to decide an election. Yet millions of us go to the trouble of voting anyway. We hold the principle of democracy in sufficiently high esteem that we've convinced ourselves, collectively, that it's our moral obligation to engage in what reason insists is a purely symbolic act.

But a vote is not a symbol. A ballot — that slip of paper — is a physical thing. We ought to regard it as a trophy, because to hold a ballot is to physically touch a profound human victory: the overthrow of all those kings and emperors who dominated so much of humanity for so much of our history. We've gotten so used to being able to choose our own government that we tend to take it for granted; in most of the world, it's not uncommon for half the electorate to stay home on election day. The consequences of that apathy can get stark in a hurry. The presidency of Donald Trump exposed that risk like nothing else, jolting Americans awake and bringing more of them to vote than in any U.S. election since 1900.

Yes, almost half of them voted for Trump. That's because everyone, even the supporters of a wannabe tyrant, intuitively grasps the strange magic that comes into force during an election: If enough of us believe our votes matter, they miraculously do. Maybe faith is a better word than magic, because the outcome of any given election truly is a matter of belief. It's worth thinking about that for a moment, especially for an atheist like me: Democracy is an exercise that translates pure belief into material reality.

Nobody knows this better than those who've been denied the right to participate. "The vote is the most powerful nonviolent tool we have," said John Lewis, the late congressman and civil rights icon, who described the right to vote as "almost sacred." One of John Lewis's ideological heirs, the Georgian legislator and activist Stacey Abrams, picked up on that thread in June 2020, when the Black Lives Matter protests were at their height and everyone was talking about how to channel outrage into policy. While that was happening, Republicans throughout America were conniving to suppress the Black vote in advance of the coming election. In an op-ed for the *New York Times*, Abrams reiterated Lewis's belief and said he hadn't gone far enough — the act of voting wasn't *almost* sacred. "As the child of ministers, I understand his hesitation to label a simple, secular act as sacred," Abrams wrote. "Voting is an act of faith. It is profound. In a democracy, it is the ultimate power. Through the vote, the poor can access financial means, the infirm can find health care support, and the burdened and heavy-laden can receive a measure of relief from a social safety net that serves all. And we are willing to go to war to defend the sacred."

Neither Lewis nor Abrams ever suggested that voting is enough on its own. You don't show up once every four years and hope for the best. But voting *is* an essential part of our social contract, both an opportunity and a responsibility, and when the difference between winning an election comes down to less than 1 percent of the vote, as it did in Georgia and other key states, voting becomes the perfect example of how every little thing we do can matter.

The US election wasn't the only example of how individual behaviour can influence collective outcomes that 2020 brought into sharp relief. Think of the struggle Lewis and Abrams embody. The fight against racism isn't one we just leave to the higher powers. Nobody says, *What difference does it make if I'm racist or not when systemic racism is so vast?* Quite the opposite. The first thing we do is the one thing in our grasp as common citizens, which is to treat our fellow humans with respect. In addition to that, absolutely, let's pursue whatever means are available to each of us, according to our station in life, to advocate for policies that put an end to discrimination. But let's start with the daily habit of indiscriminate respect.

COVID-19 also highlighted this dynamic. Containing the pandemic is an urgent matter for policy-makers and corporations alike — there's nothing I can do to make sure hospitals have enough masks, or that unemployed citizens receive financial relief, or to hasten a vaccine's creation. When the powerful shirk those primary responsibilities, it becomes outrageous for them to advocate for "personal responsibility" instead. But the idea that individual behaviour has no bearing on the pandemic's spread is equally absurd. I wear a mask; I wash my hands; I limit my personal contacts. Canada's response may not have been perfect, but our federal leadership did seem to be trying their best under extremely trying circumstances. That gave them some authority to ask us to do the same. When the second wave began to rise, it struck me as reasonable for Canada's chief public health officer, Theresa Tam, to remind Canadians in early November 2020 that "every little thing that you do helps." One week later, Tam was echoed by President-Elect Biden, who urged all Americans to wear masks until he had the power to enact the policies Trump had belittled. "Small acts add up to enormous achievements," Biden said. "It's the weight of small acts together that bend the arc of history."

If one reason for politicians to do their jobs is that it gives them the credibility to ask us to do ours, then surely the reverse is true as well. The more individuals act like they believe in something, the more pressure it puts on our leaders to act like they believe it, too.

In 1978, in the country my father would have grown up in were it not for Hitler and the Great Depression, a playwright named Václav Havel wrote an eighty-page essay called "The Power of the Powerless." He published it as illegal samizdat, the name for all subversive literature printed underground in Czechoslovakia and throughout the Soviet Bloc.

"The Power of the Powerless" laid out a psychological road map for the overthrow of what Havel called "the post-totalitarian system" in which he and some 250 million of his fellow Soviet citizens were trapped. "I do not wish to imply by the prefix 'post-' that the system is no longer totalitarian," he clarified. "On the contrary, I mean that it is totalitarian in a way fundamentally different from classical dictatorships." The difference was that the Soviet dictatorship no longer relied on force to assure everyone's good behaviour. After several generations of state control and ubiquitous propaganda, the general population had internalized Soviet ideology.

To live in this post-totalitarian system was to live "in a world of appearances trying to pass for reality," Havel wrote. To illustrate that supremacy of appearance, he put an imaginary greengrocer at the heart of his essay. This grocer, like all his neighbours, placed a sign in his window every day that read "Workers of the world, unite!" Why did he do that? Was it because he ardently believed in the common plight of workers around the world? No. "The overwhelming majority of shopkeepers never think about the slogans they put in their windows, nor do they use them to express their real opinions," Havel wrote. The grocer put that sign there "because it has been done that way for years, because everyone does it, and because that is the way it has to be." Of course, the grocer also knows that "if he were to refuse, there could be trouble," but in a post-totalitarian society that threat has receded to the background, which is the crucial point. Fear of violence had been replaced by a numb acceptance of the status quo. "By this very fact, individuals confirm the system, fulfill the system, make the system, are the system."

That system was built on a foundation of untruth: "It falsifies the past. It falsifies the present, and it falsifies the future. It falsifies statistics. It pretends not to possess an omnipotent and unprincipled police apparatus. It pretends to respect human rights. It pretends to persecute no one. It pretends to fear nothing. It pretends to pretend nothing."

Some of this sounds familiar, and not only to readers of *Nineteen Eighty-Four*. It happens to also bear a striking resemblance to the parallel reality Fox and other far-right media have been building in the United States for two decades, a system that has now ensnared tens of millions of Americans, including several of my relatives. But even those of us who find the Fox world view repugnant, who are horrified by the assault on truth and democracy that's unfolded before our eyes in the United States and elsewhere — even we are bound up in the system that produced Trump; our system, too, is based on a lie.

Our system — let's call it modern capitalism — is the one that brought us climate change and the world's sixth great extinction. Our system is the one that tried to bury its legacy of genocide and slavery beneath the prosperity that grew out of those crimes. We all have a different relationship with that system, and every day more of us are in open revolt. But given the direction that the world's ecosystems are headed, given the fact that not a single country on Earth is anywhere near to meeting its climate targets, let alone challenging the logic of a global economy built on infinite growth — given all that, it doesn't strike me as a wild exaggeration to compare *our* system's grip on *our* individual lives to the grip Communism had on Václav Havel's greengrocer. Which was, of course, the grip it had on Václav Havel himself. The critical difference being that Havel was aware of it.

Our lives may be better than that of Havel's greengrocer by almost any measure. We may be free to hang any sign we like in our windows, and write what we like without fear of reprisal. But as anyone who's tried living without email or a credit card knows, we are ensnared, too. It's a wonderful trap, and it's leading us to ruin.

In 1978, when I was two years old, Havel foresaw the risks we ran. He perceived that the willingness to live within a lie wasn't a function of Soviet ideology so much as an aspect of the human condition — not an inescapable trait, but one we ignore at our peril. "Is it not true that the far-reaching adaptability to living a lie," he asked, "has some connection with the general unwillingness of consumption-oriented people to sacrifice some material certainties for the sake of their own spiritual and moral integrity? With their vulnerability to the attractions of mass indifference? And do we not in fact stand as a kind of warning to the West, revealing to [the West] its own latent tendencies?"

Indeed you did, and four decades later here we are. For all the enormous differences between our system and Havel's, some critical similarities are worth emphasizing. Not just the way we are both ensnared in a system with deadly consequences but also the internal, psychological tension both systems produce. I have described that tension as a paradox: the uncomfortable dissonance that arises when we become aware of the incalculable damage our way of life is inflicting on this planet, and on future generations. Havel described his internal tension like this:

> In everyone there is some longing for humanity's rightful dignity, for moral integrity, for free expression of being and a sense of transcendence over the world of existence. Yet, at the same time, each person is capable, to a greater or lesser degree, of coming to terms with living within the lie. Each

person somehow succumbs to a profane trivializa-
tion of his inherent humanity, and to utilitarianism.
In everyone there is some willingness to merge with
the anonymous crowd and to flow comfortably
along with it down the river of pseudolife. This is
much more than a simple conflict between two identities. It is something far worse: it is a challenge to
the very notion of identity itself.

When I read those words, I think of the pandemic that swept
across the so-called Western world long before COVID-19 and
which will be here long after COVID is gone: anxiety, and its
sibling, depression. The World Health Organization calculates
some 300 million people around the world suffer from an anxiety disorder, including almost one in five Americans; the same
number of people report depression. These numbers are going
up, not down, as development and democracy sweep the world.
It would be facile to declare that our declining mental health is a
direct result of ecological calamity, but wouldn't it also be naive
to suppose there's no connection? We evolved, like all species, to
be in relationship with the natural world. Severing that relationship is going to have consequences, and these consequences can
only grow more severe as evidence mounts that we're incinerating the future.

There is a profound dissonance built into our daily lives. It's
not the only thing making us anxious and depressed, but it surely
plays a major role. It's not helping our politics, either.

Havel's greengrocer, too, suffered under the weight of his own
life's contradictions. But he wasn't helpless. Though he had no
hope of overthrowing the system that enveloped him, he could
take steps to liberate himself, and in so doing rattle the chains just
a little. He could take down the sign and stop pretending he cared
whether the workers of the world united or not.

A system based on a lie "works only as long as people are willing to live within the lie," Havel wrote. The moment the greengrocer took that sign out of the window, he started a micro-revolution. He "shattered the world of appearances, the fundamental pillar of the system. He has upset the power structure by tearing apart what holds it together. He has demonstrated that living a lie is living a lie."

I don't want to take the comparison too far. The consequences for a simple act of dissent in 1978 Czechoslovakia were much harsher than anything comparable in modern North America. The grocer risked losing his shop, imperiling his children's education, or being sent to jail if he persisted. What do any of us suffer if we choose to become a vegetarian, or stop using plastic bags, or cease taking our children to Disneyland? Inconvenience, raised eyebrows, heavy sighs. But shouldn't this lack of consequence serve to encourage us?

Clearly, it's not fear of repercussion stopping us. Instead, a sense of impotence has yielded apathy; that familiar fear of inconsequentiality, of knowing that nothing I do will slow climate change, has a profoundly paralyzing impact. If nothing I do makes a difference, why should I do anything at all? Why should *I* suffer the sacrifice of living a simpler life?

I submit that the chief benefit of doing whatever little things we can is personal. Becoming aware that every little thing we do has some impact, and acting accordingly gives our lives purpose. It imbues our humdrum daily routine with a little hit of meaning. To eat with intention, to reduce our consumption of material goods, to drive a little less and walk a little more, and to choose our leaders carefully — none of these things are guaranteed to change the world. But they're likely to make us feel better.

And you never know. Sometimes, the world does change as a result of these multitudinous actions.

Václav Havel believed this in 1978, the year he published "The Power of the Powerless," when there was really no reason to think so; the Soviet Union's grip on its empire appeared total, and people like Havel had never felt so powerless. And yet he had the faith to write, "It is never quite clear when the proverbial last straw will fall, or what that straw will be."

Immediately after he wrote that essay, he came under unbearable pressure from the Czech authorities. They put him under constant surveillance and interrogated him twice a day for months. Then they threw him in jail for four years. Five years after he was released, the Velvet Revolution culminated in the peaceful overthrow of Czechoslovakia's communist government. The Soviet Union was collapsing. One month later, on December 29, 1989, Václav Havel became president.

Not everyone has to risk imprisonment or run for high office. History is awash with movements that prevailed because enough greengrocers took down their signs and started talking to their neighbours. When the moment of transformation arrives, it often seems sudden as a shore-breaking wave, but in reality change was gathering beneath the surface all along, swelling imperceptibly toward its breaking point, one person at a time.

ACKNOWLEDGEMENTS

I WROTE THIS BOOK FROM my home in the unceded territories of the Squamish, Musqueam, and Tsleil-Waututh Nations. Their age-old stewardship of this stunning place, and their ongoing struggle for justice, is a permanent source of inspiration and outrage that no words of mine can sufficiently honour. I am forever in their debt.

I am also indebted to my fabulous editor, Jen Knoch, whose wisdom, patience, and passion for all things literary and ecological are just a few of the qualities that make her the perfect co-conspirator. Thanks for seeing this through with me, Jen; every writer should be so lucky.

I would never have met Jen if it weren't for my agent, Stephanie Sinclair, who believed in this project from the moment I mentioned it, and helped me believe in it, too. Thank you, Stephanie!

Huge thanks as well to Crissy Calhoun, a tremendous writer and copy editor whose falcon eyes and careful contemplation were a godsend.

In 2018 I pitched an essay about the environmentalist's paradox to Mark Medley, Opinion editor at the *Globe and Mail*. That essay became the foundation of this book. I owe you, Mark. Thanks also to Harley Rustad and Carmine Starnino at *The Walrus*, and to Kyle Wyatt at the *Literary Review of Canada*, for helping me bring about two more of the essays that shape-shifted their way into this collection ("Portents and Prophecies," and "Dangerous Opportunities," respectively).

Big thanks to my parents, Karl and Rita, who graciously answered my endless questions about their upbringing and agreed to let me share their stories with the world. That's a dangerous thing to do. In the same vein, thank you to Heather McPherson, and to all the Vancouver members of Extinction Rebellion (Lexa and Harold in particular), for putting up with me for a whole year.

I'll never forget the semester I spent in Ann Arbor as a Knight-Wallace Journalism Fellow, in 2016. That experience helped inform "The Suspension of Disbelief" and "Let's Get Drunk and Celebrate the Future." Thank you forever, Lynnette Clemetson, intrepid director and gracious host of the Knight-Wallace Fellowship, for believing in me and helping me think through an impossible subject. I must admit, I didn't think it would come out quite like this.

Speaking of unexpected outcomes: Several years ago, the Canada Council for the Arts and the B.C. Arts Council both took a chance on me, extending their generous support for what we all thought was going to be a novel. After no small anguish, this book of essays emerged instead. Let the record state, for better or worse, you don't always get what you pay for. Thank you to both institutions and the people who make them possible. I'm grateful to live in a country that supports its writers so.

Of all the people who supported me throughout the writing of this book, none deserves more credit than my darling Kiran, who has now endured three of these blasted endeavours. Somehow, she found it in her to be a bottomless well of emotional and material uplift while navigating her own harrowing year of service on the frontlines of a pandemic. Thank you, love, for the walks and the talks, the dinners and the grins, the runs in the woods, and so much more. No one knows better than you (except me) that I couldn't have done this without you.

Lastly, I must thank my daughter, Ada Jane, and ask her forgiveness: You were too young to consent to your appearance in these pages, and I can only hope that once you're old enough to read these words, you'll be all right with them. You inspired much of this book, and sustained its transition from impulse to final draft with your regular office delivery of hugs and drawings. As the months wore on, those drawings started to include your own first, carefully written words. There was one word in particular that served as a daily reminder of what it all comes down to: *love*.

SOURCES

THE NEWEST NORMAL

The forests of British Columbia have started to emit more carbon than they absorb: "B.C.'s Forests: Full Decade of Carbon Loss," Sierra Club, June 2015.

Global maternal mortality rates, pre-industrial and present: Max Roser and Hannah Ritchie, "Maternal Mortality," Our World in Data.

Child mortality rates in Canada one century ago vs. today: "Child Mortality Rate (under Five Years Old) in Canada, 1830–2020," Statistica.

Between 2000 and 2017, global maternal mortality dropped by 38 percent: "Maternal Mortality," Unicef, September 2019.

. . . while under-five child mortality was cut in half: "Mortality Rate, under-5 (per 1,000 Live-Births)," World Bank.

A study published in the *Proceedings of the National Academy of Sciences*: Yinon M. Bar-On, Rob Phillips, Ron Milo, "The Biomass Distribution on Earth," *PNAS*, June 19, 2018

A United Nations report warned that up to one million species face extinction: "Global Assessment Report on Biodiversity and Ecosystem Services," *Intergovernmental Science-Policy Platform on Biodiversity and Ecosystem Services*, May 6, 2019.

Southern Resident orcas are on the brink of extirpation, down to less than eighty individuals: see, for instance, Randy Shore, "Tougher Measures Needed to Save Southern Resident Killer Whales, Experts Warn," *Vancouver Sun*, January 23, 2021.

There are still some 700 million undernourished people in the world: "As More Go Hungry and Malnutrition Persists, Achieving Zero Hunger by 2030 in Doubt, UN Report Warns," World Health Organization, July 13, 2020.

. . . plus eighty million refugees and internally displaced peoples: "Figures at a Glance," United Nations High Commissioner for Refugees, June 18, 2020.

Over 50 percent of the kids in Canada's foster care system are Indigenous: Canada Census, 2016.

The proportion of incarcerated Americans who are Black is three times that of the general population: John Gramlich, "Black Imprisonment Rate in the U.S. Has Fallen by a Third Since 2006," Pew Research Center, May 6, 2020.

U.S. maternal mortality is up from seven deaths per 100,000 in 1990 to over seventeen today (a figure that rises to forty-two for Black women): "2018 Maternal Mortality Rate," Centers for Disease Control and Prevention.

A hundred years ago, the number was over six hundred: "Achievements in Public Health, 1900–1999: Healthier Mothers and Babies," *Morbidity and Mortality Weekly Report*, CDC, October 1, 1999.

"This is an extraordinary time": Rebecca Solnit, *Hope in the Dark*, third edition (Chicago: Haymarket, 2015).

"Remembering is a radical act": George Monbiot, "In Memoriam," *The Guardian*, June 29, 2018.

A version of this essay, "Things Have Never Been So Good for Humanity, nor So Dire for the Planet," first appeared in the *Globe and Mail* on August 25, 2018.

MICKEY MOUSE IS ALL RIGHT

In 2019, the carbon dioxide censors on Mauna Loa would surpass 410 parts per million: Global Monitoring Laboratory, National Oceanic and Atmospheric Administration.

The Walt Disney Company posted fourth-quarter [2019] earnings of $19 billion: "The Walt Disney Company Report Fourth Quarter and Full Year Earnings for Fiscal 2019," The Walt Disney Company.

Walt Disney, "What I want Disneyland to be most of all is a happy place": Dave Smith, *The Quotable Walt Disney* (New York: Disney Editions, 2001).

The Sherman brothers . . . described this one as "a prayer for peace": *The Boys: The Sherman Brothers' Story*, directed by Gregory V. Sherman and Jeffrey C. Sherman (2009).

Mickey prancing around in blackface: Zaron Burnett III, "Mickey Mouse Proves You Can't Erase the Racism of Blackface," *Mel*, 2020.

Walt Disney, "I can't believe that there are any heights that can't be scaled by a man who knows the secret of making dreams come true": Kevin A. Martin, *Perceive This! How to Get Everything You Want Out of Life by Changing Your Perceptions* (iUniverse, 2004).

The internet accounts for over 3 percent of global emissions, comparable to the global airline industry: Sarah Griffiths, "Why Your Internet Habits Are Not as Clean as You Think," BBC, March 5, 2020.

THE VELOCITY OF PERCEPTION

One of those churches was set on fire with four hundred people inside: Abdullahi Boru Halakhe provides a full account of this atrocity in his report on Kenya's post-election violence in *R2P in Practice: Ethnic Violence, Elections and Atrocity Prevention in Kenya*, Global Centre for the Responsibility to Protect, 2013.

The "spiritual calluses" that Gary Wills wrote of: Gary Wills, *The Second Civil War: Arming for Armageddon* (New York: New American Library, 1968). Wills writes, "The constant rub against hostility, a steady prickle of danger, forms spiritual calluses."

The general public can watch the virus spread through 188 countries "in real time": Ensheng Dong, Hongru Du, Lauren Gardner, "An Interactive Web-Based Dashboard to Track COVID-19 in Real Time," *The Lancet 20*, no. 5 (February 19, 2020).

A Champions League match in Milan draws forty thousand fans . . . Italy's hospitals "are like the trenches of a war": Jason Horowitz, "We Take the Dead From Morning Till Night," *New York Times*, March 27, 2020.

At the height of the lockdown, civilization's daily emissions down by 17 percent: "Temporary Reduction in Daily Global CO_2 Emissions During the COVID-19 Forced Confinement," *Nature Climate Change 10* (May 19, 2020).

In 2018, Bitcoin mining consumed as much energy as all the world's solar panels produced: David Wallace-Wells, *The Uninhabitable Earth: Life After Warming* (New York: Tim Duggan Books, 2019).

By the end of the year, global emissions will be back above pre-pandemic levels, reducing 2020's drop to roughly 6 percent: "After Steep Drop in Early 2020, Global Carbon Dioxide Emissions Have Rebounded Strongly," International Energy Agency, March 2, 2021.

Rob Jackson, "We don't want a Great Depression": CBC Radio, *The Current,* March 15, 2020.

A 2008 study published in *Nature*: Kate E. Jones, Nikkita G. Patel, Marc A. Levy, Adam Storeygard, Deborah Balk, John L. Gittleman, Peter Daszak, "Global Trends in Emerging Infectious Diseases," *Nature 451* (February 21, 2008).

Canada's Finance Minister, Chrystia Freeland, gives a speech: Address to the Toronto Global Forum, "Canada's Plan for a Strong Economic Recovery from COVID-19," October 30, 2020.

The Canadian government soon issues a report describing COVID as a "potential extinction-level event" for clean energy: Mia Rabson, "More Aid for Fossil Fuels Than Clean Energy in COVID-19 Response," Canadian Press, July 15, 2020.

Renewables provided more power to the U.S. electric grid than coal for every single day of April: Seth Feaster, "Renewables Surpass Coal in U.S. Power Generation throughout the Month of April 2020," Institute for Energy Economics and Financial Analysis, May 4, 2020.

In the U.K., solar power helped set a new record: Emily Pontecorvo, "In the Middle of a Pandemic, Renewables Are Taking over the Grid," *Grist,* May 1, 2020.

The International Energy Agency announces that solar power has become the cheapest form of electricity on Earth: "World Energy Outlook 2020," International Energy Agency, October 13, 2020.

Alberta opens fifty thousand square kilometres of the Rocky Mountains to coal mining: "Statement on Open-Pit Coal Mining in Alberta's Rockies," Yellowstone to Yukon Conservation Initiative, June 3, 2020.

In the following two weeks, over two thousand protests are held across the country . . . fifteen to twenty-six million people turned out: Larry Buchanan, Quoctrung Bui, Jugal K. Patel, "Black Lives Matter May Be the Largest Movement in U.S. History," *New York Times*, July 3, 2020.

Support for Black Lives Matter jumps from 67 percent to 78 percent of Americans: Nate Cohn and Keven Quealy, "How Public Opinion Has Moved on Black Lives Matter," *New York Times*, June 10, 2020.

General Mark Milley, "I should not have been there": Televised statement to U.S. armed service members, June 11, 2020.

The Great Lakes are warmer than they've ever been: Jason Samenow, "Great Lakes Water Temperatures Are Blowing Away Records and Could Climb Higher," *Washington Post*, July 14, 2020.

Oil and gas industry sheds 100,000 jobs: Katherine Dunn, "The Oil and Gas Industry Has Lost More Than 100,000 Jobs This Year," *Fortune*, October 4, 2020.

In the first half of 2020 the G20 spends $200 billion on fossil fuel bailouts and barely half that on clean energy: Mia Rabson, "More Aid for Fossil Fuels Than Clean Energy in COVID-19 Response," Canadian Press, July 15, 2020.

Researchers report that Greenland's ice sheets have passed the tipping point, "no longer changing in geological time": Michalea King, lead

author of study in *Communications Earth and Environment*, quoted
in EurekAlert!, "Warming Greenland Ice Sheet Passes Point of
No Return."

An electrical storm unleashes eleven thousand bolts of lightning
on California without a drop of rain: Paul Murphy, "About 11,000
Lighting Bolts Strike California, Igniting Hundreds of Fires," CNN,
August 20, 2020.

Four times as many women lose their jobs as men: Julie Kashen, Sarah
Jane Glynn, Amanda Novello, "How COVID-19 Sent Women's Workforce
Progress Backward," Center for American Progress, October 30, 2020.

The European Union will generate more power from renewables
than fossil fuels for first time in 2020: Dave Vetter, "It's Official: In
2020, Renewable Energy Beat Fossil Fuels Across Europe," *Forbes*,
January 25, 2021.

Scientists warn that 40 percent of the Amazon is poised to turn into
savannah: Arie Staal, et al., "Hysteresis of Tropical Forests in the 21st
Century," *Nature Communications 11* (October 5, 2020).

Hurricane Delta becomes the fastest storm ever to strengthen into a
Category 4: Jason Samenow, Ian Livingston, "Hurricane Delta by the
Numbers," *Washington Post*, October 12, 2020.

Climate-related disasters cost the United States $95 billion in 2020: Adam
B. Smith, "2020 U.S. Billion-Dollar Weather and Climate Disasters in
Historical Context," NOAA Climate.gov.

THE SUSPENSION OF DISBELIEF

People were calling in to report that antifa was lighting the fires:
"Oregon Officials Warn False Antifa Rumors Waste Precious
Resources For Fires," NPR, September 13, 2020.

Coleridge, "that willing suspension of disbelief": Samuel Coleridge, *Biografia Literaria* (1817).

Hannah Arendt, "The ideal subject of totalitarian rule": Hannah Arendt, *The Origins of Totalitarianism* (Schocken Books, 1951).

The Bush Administration had already drawn up plans to invade Iraq before 9/11: Jonathan Stein and Tim Dickinson, "Lie by Lie: A Timeline of How We Got into Iraq," *Mother Jones*, September 2006.

THE UNBEARABLE WHITENESS OF BEING (AN ENVIRONMENTALIST)

Over 200,000 Canadians experience homelessness each year: Stephen Gaetz, Erin Dej, Tim Richter, Melanie Redman, *The State of Homelessness in Canada 2016*, The Canadian Observatory on Homelessness, 2016.

Nearly one in five Canadian children live in poverty . . . one in three racialized children . . . 41 percent of First Nations children living off reserve, and over half (53 percent) of First Nations children living on reserve: "Beyond the Pandemic: Rising Up for a Canada Free of Poverty," Campaign 2000: End Child & Family Poverty, United Way, 2020.

Indigenous children make up over half the kids living foster care: "Reducing the Number of Indigenous Children in Care," First Nations Child and Family Services, Government of Canada, 2020.

There are roughly 2000 hate crimes committed each year across the country: "Police-Reported Hate Crime, by Type of Motivation," Statistics Canada, 2015–2019.

Black Canadians comprise nearly 9 percent of the prison population: Anthony N. Morgan, "Black Canadians and the Justice System," *Policy Options*, May 8, 2018.

Indigenous Canadians comprise 30 percent of the prison population: "Indigenous People in Federal Custody Surpasses 30%," Office of the Correctional Investigator, Government of Canada, January 21, 2020.

Robert Bullard, "America is segregated and so is pollution": drrobertbullard.com.

A 2020 report by the UN Special Rapporteur on Human Rights and Toxics: Baskut Tuncak, *Report of the Special Rapporteur on the Implications for Human Rights of the Environmentally Sound Management and Disposal of Hazardous Substances and Wastes on His Visit to Canada*, September 4, 2020.

In the Ojibwe community of Grassy Narrows . . . the Mi'kmaq community of Pictou Landing . . . Cree communities living downstream of the oil sands: Ingrid R.G. Waldron, *There's Something in the Water* (Black Point, NS: Fernwood, 2018).

In Vancouver, the Musqueam Nation has spent decades living with the intolerable stench of a sewage pipe: "Vancouver to spend $30-million on Odour Treatment Plants to Deal with Foul Sewage Smell," *National Post*, March 19, 2015.

A newly minted Nazi party began studying U.S. legislation that created the legal architecture for America's pursuit of genocide . . . one American whom Adolf Hitler particularly admired was Madison Grant: Jedediah Purdy, "Environmentalism's Racist History," *New Yorker*, August 13, 2015.

John Muir, "The clearest way into the universe is through a forest wilderness": John Muir, *John of the Mountains: The Unpublished Journals of John Muir* (Boston: 1938).

Muir described the Cherokee homes as "the uncouth . . . wigwams of savages" . . . later in his life, Muir had a change of heart: Justin Nobel, "The Miseducation of John Muir," *Atlas Obscura*, July 26, 2016.

Aldo Leopold worried that immigrants would "overrun the country": Eve Andrews, "Why Does Environmentalism Have a Dark Side?" *Grist*, August 19, 2019.

Robert Bullard, "A good first baby step": quoted by Darryl Fears and Steven Mufson, "Liberal, Progressive — and Racist? The Sierra Club Faces Its White-Supremacist History," *Washington Post*, July 22, 2020.

Aboriginal law is both the newest and fastest-evolving section of Canada's legal code . . . Meares Island, Aboriginal Title, Xeni Gwet'in: Arno Kopecky, "Title Fight," *The Walrus*, July 2015.

LET'S GET DRUNK AND CELEBRATE THE FUTURE

I have before me a number of hardcover tomes: Jeff Rubin, *The End of Growth*, 2012; Robert J. Gordon, *The Rise and Fall of American Growth*, 2016; Douglas Rushkoff, *Throwing Rocks at the Google Bus: How Growth Became the Enemy of Prosperity*, 2017; Vaclav Smil, *Growth: From Microorganisms to Megacities*, 2019.

The forty-four richest people in my country made $64 billion last year: Daniel Tencer, "Canada's Richest 44 People Add $63.5 Billion in Wealth as 20% of Low-Income Jobs Vanish," *Huffington Post*, January 26, 2021.

REBEL, REBEL

An international poll released a few weeks before the Burrard Bridge blockade: Matthew Taylor, "Climate Crisis Seen as 'Most Important Issue' by Public, Poll Shows," *The Guardian*, September 18, 2019.

Christiana Figueres, "civil disobedience is not only a moral choice": Christiana Figueres and Tom Rivett-Carnac, *The Future We Choose: Surviving the Climate Crisis* (New York: Knopf, 2020).

Her name is Lexa: Everything written here about Lexa is done with her full knowledge and consent.

INSIDE JOB

Alberta was earning more from cannabis and liquor sales than it was from oil and gas: Gordon Jaremko, "Drinking and Gambling Outpace Oil, NatGas Revenues in Canada's Top-Producing Province," Natural Gas Intelligence, March 1, 2019.

United Conservative Party forced to cancel celebration of carbon tax cancellation because of smoke from forest fires: Nadine Yousif, "Kenney Cancels Carbon Tax Repeal Celebration Due to Ominous Wildfire Smoke," Toronto Star, May 30, 2019.

Husky, Cenovus, Suncor, and other fossil fuel behemoths lay off a quarter of their workforce: John Gibson, "Cenovus to Cut up to 25% of Combined Workforce with Husky Energy after Merger," CBC News, October 27, 2020.

The party that based its 2020 budget on oil prices of $58 per barrel: Chris Epp, "Oil Price Plunge Puts Alberta Budget into Question," CTV News, March 6, 2020.

The party that made protesting energy infrastructure a criminal offence: Taylor Lambert, "The Truth Behind the UCP's Anti-Protest Law," The Sprawl, June 20, 2020.

The party that created an information "war room" . . . to publicly display a spectacular incompetence: Andrew Leach, "The Fiasco That Is Alberta's Energy 'War Room,'" CBC News, July 30, 2020.

Devin Dreeshen spent much of 2015 volunteering for the Trump campaign: Hadeel Abdel-Nabi, "Conservative Candidate in Alberta Hiding Past as a Trump Campaigner," Vice, July 11, 2018.

Kenney's speech writer, "everyone knows that race is the defining element of violent crime in Canada": Ashley Joannou, "NDP Renews

Calls for Kenney Speechwriter to Be Fired over Controversial Articles,"
Edmonton Journal, July 1, 2020.

The UCP considered removing the history of residential schools from
elementary curriculum: Janet French, "Education Experts Slam Leaked
Alberta Curriculum Proposals," CBC News, October 21, 2020.

The education minister's chief of staff was caught posing with
Soldiers of Odin: Marianne Maravilla, "UCP Candidates under Fire
for Posing with Soldiers of Odin Members at Event," CTV News,
October 7, 2018.

The member from Lac Ste. Anne complained that federal relief
cheques were "funny money": Lisa Johnson, "'They make more
on CERB, eating Cheezies': UCP MLA who said federal Program
Encourages People to Stay Home, Fuels Drug Use Says Comments
Taken Out of Context," *Edmonton Journal*, September 29, 2020.

The "Inquiry into Un-Albertan Activities" produced nothing but
unintended irony: Martin Olszynski, "'Textbook Climate Denialism': A
Submission to the Public Inquiry into Anti-Alberta Energy Campaigns,"
The University of Calgary Faculty of Law Blog, January 14, 2021.

Jason Kenney, "a premeditated, internationally planned and financed
operation to put Alberta out of business": quoted by Tyler Dawson,
"Alberta Announces Public Inquiry into 'Shadowy' Foreign Funding of
Environmental Groups," *National Post*, July 4, 2019.

A poll released one week after Aloha-gate: "Poll Shows Sharp Decline
in UCP Support as NDP and Wildrose Gain Ground," CTV News,
January 9, 2021.

The UCP had just sold a few of the nearby mountains to an Australian
coal magnate: Kim Siever, "Alberta Awards 11 Coal Leases in SW Alberta
for $36 a Hectare," Kim Siever News, December 18, 2020.

A Greenpeace organizer from Edmonton who later published a story about the experience: Mike Hudema, "I Spent over 36 Hours Suspended from a Vancouver Bridge to Stop a Tar Sands Tanker," Greenpeace.org, July 9, 2018.

PORTENTS AND PROPHECIES

The United States could cut 70 percent of its emissions just by electrifying everything . . . doing so would create at least fifteen million jobs: David Roberts, "How to Drive Fossil Fuels Out of the US Economy, Quickly," *Vox*, August 6, 2020.

We're dumping nearly forty gigatons of carbon dioxide into the atmosphere each year: Zeke Hausfather, "Global Fossil-Fuel Emissions up 0.6% in 2019 Due to China," *Carbon Brief*, December 4, 2019.

California used modern capitalism to lower emissions: Mark Jaccard, "I Wish This Changed Everything," *Literary Review of Canada*, November 2014.

A BRIEF HISTORY OF POPULATION CONTROL

Theodore Roosevelt, "race suicide": Adam Hochman, "Race Suicide," EugenicsArchive.ca, April 29, 2014.

Winston Churchill, "60,000 imbeciles, epileptics and feeble-minded": Martin Empson, "Dispelling 'the Malthus Myth,'" *International Socialism*, June 25, 2010.

Through the 1970s, some 1,200 Indigenous women were sterilized against their will: Roger Collier, "Reports of Coerced Sterilization of Indigenous Women in Canada Mirrors Shameful Past," *CMAJ News*, August 2, 2017.

Inuit communities saw up to a quarter of their adult women sterilized . . . in 2019, over one hundred Indigenous women from six provinces signed onto a class action lawsuit alleging they'd been sterilized between 1985 and 2018: Avery Zingel, "Indigenous Women Come Forward with Accounts of Forced Sterilization, Says Lawyer," CBC News, April 18, 2019.

A quarter of the world's groundwater extraction occurs in India: Neha Jain, "India's Groundwater Crisis, Fueled by Intense Pumping, Needs Urgent Management," *Mongabay*, June 7, 2018.

China down to its last fifty tigers: "13 Countries Where You Might Find Wild Tigers," World Wildlife Fund, July 27, 2018.

An estimated thirty million more marriageable men than women in China: Kristal Sotamayor, "The One-Child Policy Legacy on Women and Relationships in China," PBS, February 5, 2020.

China leads the world in carbon emissions, twice as much as the U.S. and four times as much as India: "Each Country's Share of CO_2 Emissions," Union of Concerned Scientists, August 12, 2020.

As a result of women's empowerment, the world's population is settling down: Monica Das Gupta, "Women's Empowerment and Fertility: Policy Lessons," United Nations Department of Economic and Social Affairs, 2013.

DANGEROUS OPPORTUNITIES

Aladin El-Mafaalani discussed *The Integration Paradox* in a speech at the Canadian embassy in Berlin: "Fighting at the Table: Conflict as Successful Integration," *CBC Ideas*, June 29, 2017.

ONCE UPON A TIME IN DEUTSCHLAND

This was the mid-1980s, when the two Germanys were at the height of atoning for the Holocaust: Susan Neiman, *Learning from the Germans: Race and the Memory of Evil* (New York : Farrar, Straus and Giroux, 2019).

The biggest single act of ethnic cleansing in history, fourteen million ethnic Germans from eastern Europe: R.M. Douglas, *Orderly and Humane: The Expulsion of the Germans after the Second World War* (New Haven: Yale University Press, 2012).

Rising-tide economics converged with trickle-down theory to soak three-quarters of the population: Rakesh Kochhar, "The American Middle Class Is Stable in Size, but Losing Ground Financially to Upper-Income Families," Pew Research, September 6, 2018.

History of U.S. top marginal tax rate: "Historical Highest Marginal Income Tax Rates," Tax Policy Center, Urban Institute & Brookings Institution.

Francis Fukuyama could, in 1992, write a book arguing the end of history had arrived: Francis Fukuyama, *The End of History and the Last Man* (New York: The Free Press, 1992).

40 percent of the ice in Glacier National Park, gone: Scottie Andrew, "Some of Glacier National Park's Glaciers Have Lost as Much as 80% of Their Size in the Last 50 Years," CNN, September 16, 2020

Two-thirds of the world's rivers are dammed: Stefan Lovgren, "Two-Thirds of the Longest Rivers No Longer Flow Freely — and It's Harming Us," *National Geographic*, May 8, 2019.

Thomas Mann, "It would be hard to find a politician": Thomas Mann, "Germany and the Germans," public address delivered in the Library of Congress, May 29, 1945.

Grass's magical realism inspired Salman Rushdie: Salman Rushdie, "On Günter Grass," *Granta*, March 1, 1985.

Grass, "Born in 1927, I belong to a generation": Günter Grass, "By a Rough Estimate," *The Günter Grass Reader*, trans. Helmut Frielinghaus (Orlando: Harcourt, 2004).

Grass, "Take the goose step out of German": Günter Grass, "To Be Continued . . ." Nobel Lecture, 1999.

Rushdie, "tore their language down and rebuilt it anew": Salman Rushdie, "Günter Grass," *Imaginary Homelands: Essays and Criticism 1981–1991* (London: Granta, 1991).

Grass, "a speechless event that left nothing unsaid": Günter Grass, "Willy Brandt at the Warsaw Ghetto," *The Günter Grass Reader*, trans. Helmut Frielinghaus, (Orlando: Harcourt, 2004).

Grass, "Writers have recently begun to take an interest": Günter Grass, "Writers and the Trade Unions," *On Writing and Politics 1967–1983*, trans. Ralph Manheim (San Diego: Harcourt Brace,1984).

One of the greatest contributions to humanity of this guilt-ridden nation would be . . . solar power: Jillian Weinberger, Amy Drozdowska, Byrd Pinkerton, "How Germany Helped Make Renewable Energy Cheap for the Rest of the World," *Vox*, January 22, 2020.

Mann, "It was a militant slave mentality": Thomas Mann, Germany and the German (Washington: Library of Congress, 1945).

In 1940 Canada, 100 percent of a corporation's "excess wealth" was taxed: Seth Klein, *A Good War: Mobilizing Canada for a Climate Emergency* (Toronto: ECW Press, 2020).

EVERY LITTLE THING

Shell spent US$49 million in 2019 lobbying governments not to enact climate legislation: "Big Oil's Real Agenda on Climate Change," InfluenceMap.org, March 2019.

BP introduced the very notion of a carbon footprint: Imogen Learmonth, "How the 'Carbon Footprint' Originated as a PR Campaign for Big Oil," *thred*, September 23, 2020.

ARNO KOPECKY is an environmental journalist and author whose dispatches from four continents have appeared in the *Globe and Mail*, *The Walrus*, the *Literary Review of Canada*, *Reader's Digest*, and other publications. His last book, *The Oil Man and the Sea*, chronicled the battle to keep oil tankers out of British Columbia's Great Bear Rainforest and was shortlisted for the 2014 Governor General's Award. He lives in Vancouver, British Columbia.

For every book sold, 1% of the cover price will be donated to Ancient Forest Alliance, a non-profit working to protect B.C.'s endangered old-growth forests and to ensure a sustainable, value-added, second-growth forest industry.

This book is also available as a Global Certified Accessible™ (GCA) ebook. ECW Press's ebooks are screen reader friendly and are built to meet the needs of those who are unable to read standard print due to blindness, low vision, dyslexia, or a physical disability.

Purchase the print edition and receive the eBook free!
Just send an email to ebook@ecwpress.com and include:

- the book title
- the name of the store where you purchased it
- your receipt number
- your preference of file type: PDF or ePub

A real person will respond to your email with your eBook attached. And thanks for supporting an independently owned Canadian publisher with your purchase!

Printed on Rolland Enviro.
This paper contains 100% post-consumer fiber, is manufactured using renewable energy - Biogas and processed chlorine free.